DESIGNER DOGS

DESIGNER DOGS *an exposé*

Inside the Criminal
Underworld of Crossbreeding

MADELINE BERNSTEIN

Foreword by Dr. Phil McGraw

APOLLO
PUBLISHERS

Designer Dogs: An Exposé

Inside the Criminal Underworld of Crossbreeding

Copyright © 2018 by Madeline Bernstein.

Apollo Publishers books may be purchased for educational, business, or sales promotional use. Special editions may be made available upon request. For details, contact Apollo Publishers at info@apollopublishers.com.

Visit our website at www.apollopublishers.com.

Library of Congress Cataloging-in-Publication Data is available on file.

Cover design by Rain Saukas.

Print ISBN: 978-1-94806-206-0
Ebook ISBN: 978-194806-213-8

Printed in the United States of America.

To shelter pets everywhere and to those
who selflessly care for them.

CONTENTS

BY DR. PHIL MCGRAW

I have spent most of my life, actually since I was twelve years old, focusing on and studying *why people do what they do and don't do what they don't do.* Think about that concept for a minute. If you have real insight and understanding into what motivates people, you have the ability to influence their thoughts and behaviors—not in some nefarious evil way, but just by *talking about things that matter to people who care.*

That is exactly what Madeline Bernstein does in her ground-breaking, "call to action" book, *Designer Dogs.* Madeline talks about things that matter to those of us who care, and we all should. We can all have a role, whether passive or active, in the lives of millions of innocent and loving puppies born in America every year.

I jumped at the chance to write this foreword because I want to challenge everyone who picks up this book to ask himself or herself, and implore friends and family to ask themselves, *why* they are acquiring or considering acquiring a dog. There are right reasons and wrong reasons. And your selection of *which dog* to add to your family is just as important as the decision to get a dog at all.

Be honest with yourself, weigh the decision carefully. If you are acquiring a dog as you would an item of jewelry or to complete your macho image, and you are making your selection based on what is in at the moment because of some movie or because some person you admire has one, here is some advice: don't do it.

Furthermore, if the dog you think you want is some high-priced designer dog, then acquiring that dog is nothing short of an act of cruelty and abuse. That dog has been unnaturally forced into existence and is very likely unhealthy and unhappy. For every designer dog purchased, a mutt in need of adoption is left in a cage wishing someone would scruff up his ears and take him home.

Thirteen years ago, I had the privilege of adopting my "shadow," Maggie. She was an eight-week-old "snowball," thought to be a Lab/husky mix. spcaLA found her under a foundation at a commercial building site. I think she grew up more Jindo than anything, but it made no difference. When I picked her up and we got in the car, she climbed up on my shoulder and wrapped herself around my neck—and my heart!

For the last thirteen years Maggie has devoted her entire existence on this earth to being by my side. If I go upstairs, she goes upstairs. If I go to the studio, she goes to the studio. If I travel to Europe, you guessed it, she goes to Europe. That's all she asks, just to hang around where I am. She is the gentlest spirit you could ever imagine, until she perceives a threat to anyone in our family. You don't want to be that threat! She would fight Bigfoot if he had a chainsaw! Just little Maggie, the mutt.

She won't fit in a teacup. She has never been in a movie or a TV series. She can't seem to get both ears going in the same direction at any given time, but she is the best dog to ever stand on four paws. You cannot get her strength of character or commitment mass made in some puppy mill or by some profiteering dog designer that overbreeds in a way that shows he has a weak constitution.

I said earlier that I am fascinated by why people do what they do and don't do what they don't do. I asked and answered that question before I contacted the amazing and devoted Madeline Bernstein and her organization, spcaLA, and found Maggie. My "why" was that I wanted to give an orphaned dog a safe and loving home and I wanted a companion to hang out with. I had no idea at the time that I was rescuing my best friend ever.

When you finish reading this insightful exposé, I pray that you will turn your back on the overbred and unsafe designer dog breeders appealing to the in crowd and go get yourself a real friend, one that needs a home.

The world works on supply and demand. When you read what Madeline has to say, you can help dry up the demand by simply not being a buyer. When you finish reading *Designer Dogs*, talk about it, recommend it, tweet and post on Facebook about it. Let's help Madeline make some noise!

Oh, and when you adopt your Maggie, pay the fees for a few more adoptions if you can afford to, so some families without the funds might have a furry friend too.

Thank you, Madeline. Thank you for caring and never stopping the fight.

IT STARTED HERE

I still hope to be able to make something out of myself, but who can do anything after Beethoven?

—FRANZ SCHUBERT

When I was a kid, I wanted two things: a piano and a collie named "Lassie." Unfortunately, my parents were just starting out and had zero disposable income. So my father made me a cardboard piano on which I could practice my lessons. In other words, I could hum while my fingers developed muscle memory for scales.

The dog was more complex. Every one of my suburban friends had a purebred collie named "Lassie," or "Laddie" if the dog was male. I knew what I wanted: Lassie, not Laddie. *Lassie.* It had to be Lassie. According to the television show, Lassie understood everything and could do anything.

One day, when I was about eight years old, my father came home with a black and brown puppy, which he'd received from a law client who had a litter of them but no funds to pay my father's fee. "Wow," I thought, "is this what collies look like when they're babies?" When I realized that it wasn't the dog that I wanted, I was horrified. Clearly my friends had parents who loved them more. Did no one care *if I fell into a well?* I was the neighborhood pariah with a black and tan mutt and no hope of rescue from trouble. I felt pretty pathetic—a cardboard piano and a mutt in a world of baby grands and collies. I was just lucky I didn't have acne.

I did not understand the high price tag and maintenance expenses necessary for a purebred collie, nor the dangers of succumbing to peer pressure. I didn't know the sad reality of how the high demand for collies contributed to setbacks for the breed, as unscrupulous breeders produced puppies at a rate unsafe for their genetic survival. Instead, I understood that I had a knockoff when other kids had the real thing, and I could only sing myself a sad song on my cardboard piano.

Of course, my mutt, Lucky, became my soul mate and was with me for eighteen years. During that time, things changed. I was considered luckier than my city friends, who weren't allowed a dog at all, and I stood out as a unique contrarian in a neighborhood of dog owners with upscale, snooty collies. In fact, many of my wealthy Manhattan classmates, who lived in apartments with prohibitions against pets, thought I was the richest of them all, with my real dog in my real backyard. Conversely, my friends in our then–lower-middle-class Yonkers neighborhood thought I was poor and unlucky, as my parents saved up for a piano instead of a collie.

The appetite for designer dogs has not abated, and the steps to acquire and engineer them have increased exponentially. Unfortunately, this is not good news for the dogs. Years later, I would become president of the Society for the Prevention of Cruelty to Animals Los Angeles (spcaLA), an organization that is dedicated to the prevention of cruelty to animals through education, law enforcement, intervention, and advocacy.

spcaLA opened its doors as both an SPCA (Society for the Prevention of Cruelty to Animals) and SPCC (Society for the Prevention of Cruelty to Children) in 1877, a time when women, children, and animals were all legally classified as property. The need for an SPCC was extreme as there were not yet child labor laws, departments of social services, or the concern to protect children from abuse. Every SPCA in the country is legally separate from the others and functions independently. There is no national umbrella organization, no chapters, and, therefore, spcaLA works completely autonomously.

As president of spcaLA, it has been my vision, pleasure, and pride to connect spcaLA's roots to the twenty-first century and deliver programs and services that protect our most vulnerable and prevent their exploitation. We are the only SPCA that has a suite of programs to assist victims of domestic violence and support at-risk youth. We're also the only one with a court diversion program for juveniles convicted of animal cruelty and/or bullying behavior that sees the connective tissue of abuse and works to remedy it.

I am frequently asked why I work in the animal welfare field, why I don't exclusively focus on the welfare of people, and when I am planning on getting a real job. My work before animal welfare was as a prosecutor and then as a deputy inspector general. Both jobs involved prosecuting and investigating crimes. In my new role, I am still investigating and prosecuting criminals. It is the same job I have always had—just a different victim. I never changed jobs!

In fact, when I began in animal welfare over twenty years ago, I saw many of the same defendants I'd seen before. Only then did I begin to truly understand the interplay between crimes against people and crimes against animals. It is all one cycle of violence. I shouted this discovery to everyone I knew, with no result. My father said my message to the world was essentially the same as a lost fart in a breeze. Now, after all these years it is generally accepted in the public square that acts of animal cruelty are frequently a precursor to or predictive of violent behavior toward people. Just watch any television crime show for that to be asserted. It is also true that often an abused animal is found or threatened in families where domestic violence is common, and that these crimes don't occur in a vacuum but rather in concert with other criminal activities. Recently the FBI has begun keeping track of crimes against animals in order to study these connections more fully.

We have all been conditioned to accept as normal the substandard treatment of animals in pet shops, zoos, circuses, pony rides, theme parks, and films, and by breeders. We see a cute puppy and not the journey taken by that puppy to get to us. We are desensitized to it and push this desensitization forward by allowing the next generation to see the same. We truly are ignorant of the high cost of cute. There is also a dearth of real statistics and accurate numbers for anything in this business, since there is no mandated reporting, no uniform statistical categories, and multitudes of animals who exist under the radar. They live on the street or in private compounds and are exploited to facilitate criminal activities. They're essentially a shadow population.

Shining a light on and stopping the maltreatment and violence inherent in the creation of designer dogs and the pain endured by those dogs intended to be cute fits squarely within my soapbox of issues and spcaLA's mission. It is my honor, now as a grown-up, to talk to you about this issue.

CHAPTER 1

THE DESIGNER DOG: Name-Brand Purebreds and Custom-Designed Dogs

I thought that nature was enough, till human nature came.

—EMILY DICKINSON

There are two types of designer dogs: name-brand purebreds and custom-designed dogs. A name-brand purebred is a dog whose lineage can be verified and is sold with the assurance that he has no family history of "contamination" with a different breed. It's widely believed that breeding dogs carefully over many generations, in a continuous cycle of purebred parents and purebred puppies, results in predictability in appearance and behavior—that is, they "breed true" and will look and act a certain way. When sold to a new owner, a name-brand purebred may come with a certification of purity from a recognized association, such as the American Kennel Club (AKC). A breeder may sell these dogs at an especially high price.

There are hierarchies in the world of name-brand purebreds that define the value of a dog. This value is determined by things such as the rarity of the breed; its success in dog shows; and the media attention it has attracted, usually because a dog of the breed has been in a popular television show or movie, or a celebrity owns and has been photographed with a dog of the breed. Almost everyone in the United States has heard of a German shepherd, but not an Azawakh, a long-legged hunting dog native to Burkina Faso, Mali, and Niger, that is rare and hard to acquire in the United States, traits that make them particularly valuable. The dalmatian is an example of a breed that became highly desirable after a film popularized it—*101 Dalmations*. The Chihuahua became increasingly in demand after starring in commercials for Taco Bell and films such as *Beverly Hills Chihuahua* and *Legally Blonde*, and becoming a popular dog for celebrities to own. Photographs of Paris Hilton and her Chihuahua Tinkerbell were a tabloid fixture throughout Hilton's rise to fame. Presidential dog breeds are eternally popular. Demand for the Portuguese water dog skyrocketed after President Barack Obama and his family adopted two dogs of the breed, Sunny and Bo. Stars of the Westminster Kennel Club Dog Show and the Crufts dog show have made other breeds, such as the standard poodle, boxer, Airedale terrier, and American and English cocker spaniels, valuable as well.

Included in the name-brand category are toy breeds. Toy breeds are a subset of an established breed that has been bred down in size to create a lapdog version of the breed. As in the case of larger purebreds, toy breed purebreds must meet basic standards and criteria for things such as coat, weight, and overall appearance to be properly authenticated by a kennel club. Not every breed has a toy category recognized by respectable kennel clubs. Chihuahuas do not, for example, but poodles do. With careful breeding by ethical breeders who value health over appearance, these small versions can be as healthy as well-bred large versions.

Toy breeds should not be confused with "teacup dogs." Teacup dogs are tiny dogs who result from the breeding of two very small parent dogs, with the goal of making the offspring as small as possible—sometimes even less than one pound. Their tiny size imposes a wealth of health risks, and no respectable kennel club recognizes them as a subset of a breed because it doesn't want to encourage their breeding. For toy breeds, size, weight, health considerations, and other affects are standardized by the industry. There are no such standards in the teacup world.

Those of us in the animal welfare world consider a breeder "legitimate," meaning high-quality and respectable, if he or she adheres to standards, guidelines, and protocols that safeguard the health and well-being of the dogs and promote the likelihood that puppies will be born healthy and their mothers will be kept healthy during childbirth. Many legitimate breeders specialize in one breed of dog and only own and sell this particular breed. Legitimate breeders only use healthy dogs for breeding, do not breed teacup dogs, and ensure gene pool diversity by not over-breeding the same dog or inbreeding by the forced mating of family members.

With the advent of DNA testing, genetic problems can be detected and avoided instead of being passed from parent to puppy. Such genetic tests cost from $75 to $300 and can help breeders determine the likelihood of such things as spinal cord disease, red blood cell defects (causes of anemia and liver failure), and hip problems. The tests provide important information to breeders, so they know which dogs to breed and which not to breed. This gives comfort to the buyers and a better quality of life to the dogs. The cost of the test can be packaged into the price the consumer pays for the dog. The DNA kits, which are regularly used by legitimate breeders, are easy to purchase and the tests are simple to run.

People whom members of the animal welfare community refer to as "illegitimate breeders" are rarely concerned with genetic testing. They also often breed by inbreeding. Because it doesn't require the acquisition of dogs from other breeders, inbreeding is considered to be a lazy and cheap way to create offspring. It's particularly popular with illegitimate breeders of purebreds because by only breeding dogs they already own, they can guarantee there's no "contamination" from another breed. Inbreeding is extremely dangerous because it eliminates or severely shrinks, depending on how closely related the mating animals are, genetic diversity, which is required to avoid passing genetic diseases and disorders to progeny. The demand for dalmatians created by the *101 Dalmatians* films led to indiscriminate inbreeding, which resulted in generations of dalmatians being born deaf and/or with a bladder disease.

A human example of the consequences of inbreeding that I learned in school and have never forgotten is the story of hemophilia and the British royal family, often called "Queen Victoria's curse" or the "royal disease." In the nineteenth century, Queen Victoria's family engaged in a practice common among royals from all over the world, only marrying other royals, that is, relatives sleeping with relatives. In Queen Victoria's family, it led to the spread of a genetic disease, hemophilia, which is a blood clotting disorder carried in a recessive gene that can become dominant when two carriers have children. This led to the untimely deaths of several family members.

Because breeders are deciding which dogs should mate, there has to be diversity and care taken to avoid the likely passing of genetic diseases and disorders. There aren't enough legitimate breeders to satisfy the demand for name-brand purebreds, and the public responds by buying them, knowingly or unknowingly, from illegitimate breeders. What compounds this problem is that even when people are determined to only buy from legitimate breeders, they are often deceived into buying from illegitimate ones after seeing a falsified paper of authentication from a kennel club.

As poorly bred dogs flood the market, they breed with other poorly bred dogs and their offspring come out even worse off than their parents. Entire breeds could be ruined this way and a point may come when it's too late to "unbreed" defects and breeds will be permanently damaged.

Custom-designed dogs are the second type of designer dogs. Custom-designed dogs are the result of a breeder mating dogs of different breeds to create puppies who have characteristics of each parent's breed. In their engineering, the breeder has certain looks and behaviors in mind for the offspring. Custom-designed dogs are bred at a buyer's request or because a breeder expects that the mixed-breed puppies will be in high-demand—in other words, profitable.

Some proponents of custom-designed dogs argue that these dogs are healthier than other dogs because their gene pools have more diversity. Another argument is that custom-designed dogs can be bred to have qualities favorable to humans, such as being hypoallergenic. There is a misbelief that moral and knowledge-able breeders can guarantee healthy custom-designed dogs with the desired results.

In fact, however, creating a custom-designed dog is an exper-iment and the results can't be guaranteed. Legitimate breeders will use two well-bred purebreds, but, even then, the results are unpredictable. If two hybrid dogs are paired, their progeny is even less predictable. If breeders intend to create a "custom-designed breed" and not just make a one-off design, the process is open to a greater number of potential problems because multiple dogs are involved in the process. Creating a new breed requires the ongoing breeding of two parent breeds. This excessive birthing can be dangerous to mother dogs and to all puppies resulting from the breeder's experiment. There's a high probability that the puppies will have genetic deformities as a result of the mixing of their parents' genes.

Incredibly, I've heard purchasers of dogs from "custom-designed breeds" brag that they have a purebred, a purebred goldendoodle, for example, which is a cross between a golden retriever and a poodle. There is no such thing as a purebred mixed-breed dog. Whenever I try to set them straight, I've been met with fierce resistance. Some people even reference certificates of authenticity they've received, as if these are evidence of a purebred. I think the fervent defense of ownership of a "purebred mixed-breed dog" and the claims that custom-made dogs are better than other dogs are the result of a custom-design mania. Once celebrities are photographed with custom-designed dogs, public desire for similar dogs soars.

The demand has been met with a flood of "knockoff" dogs. As with purebreds, legitimate breeders aren't able to satisfy the demand for custom-made dogs. Ethical and responsible breeding is a slow process, and well-educated, credentialed breeders are careful and loyal to the integrity of the breeds they produce. Illegitimate breeders, such as the operators of puppy mills (a common term for high-volume, low-quality breeding places), use less scrupulous practices. They overbreed and crossbreed to satisfy demand. They're negligent, and their dogs are given poor-quality shelter, food, and veterinary care, or sometimes no care at all. The puppies they breed are frequently damned to a life of physical and mental illnesses from genetic flaws and poor upbringing.

Most people don't know they're buying designer dog knock-offs, and others don't care. As with anything designer—such as purses, shoes, watches, and luggage—for every person who has an original, there are countless others who have fakes. People buy designer items because they like the look of the item, but they also buy them to look a certain way to others. For many people, it's important to look the part, whether you are a genuine billionaire or an imposter.

Custom-designed dogs from legitimate breeders and knock-offs are pricey. So pricey, in fact, that their cost often exceeds the combined cost of their parents, even if the parents are purebreds. In 2017, TMZ Sports reported that NBA star Stephen Curry paid $3,800 for a goldendoodle because the puppy had green eyes like Curry's wife. I wonder if the seller told him that a dog's eyes change color with age! Curry can afford a pricey dog, but other people determined to own a custom-designed dog may enter into predatory financial arrangements with disastrous consequences, having to pay both the purchase price and care costs for the dog.

CHAPTER 2

DAWN OF THE FRANKENDOG

This job of playing God is a little too big for me. Nevertheless, someone has to do it, so I'll try my best to fake it.

—LARRY WALL

The Allegory of the Labradoodle

The custom-designed dog craze began in the 1980s with Wally Conron and the labradoodle. Conron was a puppy-breeding manager for the Royal Guide Dog Association of Australia when a Hawaiian couple asked if he had any dogs who would help the woman, who was suffering from vision problems, and that the husband, who was usually allergic to dogs, could tolerate.

Conron was determined to solve this dilemma and hypothesized that if he crossbred a Labrador, a common guide dog breed, with a poodle, whose coat is less irritating for allergy sufferers, he could provide a dog that would accommodate the needs of both the wife and the husband. After he mated the dogs and the mother gave birth to five puppies, he waited until the puppies were five months old and then collected hair and saliva samples from them and sent them to the couple. The husband had no allergic reaction to the samples from one of the five puppies; the samples from the other four caused allergic reactions. Conron then repeated the experiment. Of the samples from the next litter, ten puppies, three did not irritate the man's allergies; samples from the other seven did.

Conron gave one of the dogs whose samples did not irritate the man's allergies to the guide dog association to train for the woman, but they refused, saying they only wanted to work with purebreds. So, as a marketing gimmick, Conron announced that the dog was a purebred of a new breed he had created. He called it the "labradoodle."

The story of Wally Conron and the labradoodle is disturbing on several levels. First, why did the guide dog association refuse to train a dog simply because the dog was not a pedigree? Second, why did Conron assume that any one of the dogs could be a good match for the couple? It needed to be proven that the selected dog would not ignite the man's allergy and would also be a good guide dog for the woman. Guide dogs are part of a unique pairing, rather than a mass-produced, any dog will do pairing. Third, the experiment involved seventeen dogs and what the crossbred puppies would be like was unknown. Conron could have caused fifteen dogs to be born with birth defects that would lead to a lifetime of health issues. The mother dog could have suffered or died giving birth. And then, of course, what would happen with all the other puppies? Conron planned for the couple to take one puppy, but there were no planned homes for the others. If they had been born with health problems, it would have been even harder to find them homes.

Conron's decision to mastermind a labradoodle haunted him for years. Once his invented breed became publicly known, demand for it exploded. And this is despite the fact that, as Conron learned, labradoodles don't breed true. Their coats can differ, their behavior is unpredictable, and most aren't hypoallergenic.

In the years since Conron's experiment, he has frequently admitted that a mixed-breed dog cannot be considered a purebred and that he only called his puppies members of a new breed because he wanted the Royal Guide Dog Association to train one of the dogs to help the woman suffering from vision problems. In 2014, the year mixed-breed dogs were admitted into the Westminster Kennel Club show for the first time, he was particularly outspoken about his mistake.

In an interview published by *Psychology Today* on April 1, 2014 ("A Designer Dog-Maker Regrets His Creation" by Stanley Coren), Conron said:

> I opened a Pandora's box, that's what I did. I released a
> Frankenstein. So many people are just breeding for the money.
> So many of these dogs have physical problems, and a lot of them
> are just crazy. . . .
>
> Today I am internationally credited as the first person to breed
> the labradoodle. People ask me, "Aren't you proud of yourself?"
> I tell them "No! Not in the slightest." I've done so much harm
> to pure breeding and made so many charlatans quite rich. I
> wonder, in my retirement, whether we bred a designer dog—or a
> disaster!

The story of Wally Conron and the labradoodle is an allegory of experiments and consequences. And though it's not my mission to ease Conron's conscience, it must be said that he wasn't the first person to mix breeds to create a third breed, or a variant of a parent breed. It's thought that the first designer dog in the modern history of the United States was born in the 1950s. It was the child of a cocker spaniel and a toy or miniature poodle, called a "cockapoo" or "spoodle." Information about this breed's inventor, birth location, and the impetus behind its creation is hard to reliably locate, but it does help flesh out the timeline of the custom-designed dog industry.

Celebrity owners of labradoodles include Jennifer Aniston, Barbara Eden, Christie Brinkley, Henry Winkler, and Tiger Woods, and, as usual, what celebrities have, the public wants. People pay thousands of dollars for labradoodles. Many purchasers assume they're getting a dog that will be hypoallergenic, but find themselves sneezing when the dog comes close.

With the demand for labradoodles came the usual response from money-hungry puppy mill operators: careless breeding and ample supply. Puppy mill operators also mix a wealth of other breeds with poodles. New breeds ending with "oodle" are born left and right. Each time, two parent dogs are exploited for the experiment and offspring at high risk of genetic diseases are born. The parent dogs suffer, the puppies could be sentenced to a lifetime of suffering, possibly with short life spans, purchasers end up with expensive medical bills for their dogs, and shelters end up with abandoned dogs and must decide whether to keep them alive or euthanize them, a sad practice often performed when a home cannot quickly be found for an animal. The government does not keep tabs on how many animals are euthanized each year, but it's estimated that approximately 1.5 million shelter animals are euthanized in the United States each year.[1]

---

1   Erin Greenwald, "Millions of Dogs Need Homes. Why Is It So Hard to Adopt One?" *Washington Post* (February 2, 2018), https://www.washingtonpost.com/news/ animalia/wp/2018/02/02/millions-of-dogs-need-homes-why-is-it-sometimes-hard-to-adopt-one/?utm_term=.72434160b9d8.

The Pit Bull: A Cautionary Tale

Featured in the television show *The Little Rascals*, Buster Brown shoes advertisements, and the film *Fame*, the pit bull was once and is still, if carefully chosen, considered a good family pet. Pit bulls were companions to President Theodore Roosevelt and Helen Keller, and were once viewed as "nanny" dogs—loving, loyal, and wonderful with children. The love for this breed began to wane about twenty-five years ago as its reputation was degraded in response to news reports of dogfighting and vicious mauling incidents.

The bulldog is one of the parent dogs of the pit bull. Beginning in England in the 1600s, the bulldog was used to hunt boar, herd sheep, and participate in bullbaiting and bearbaiting contests, a blood sport extant since medieval times that involved pitting a captive bear or bull, confined in a ring, against dogs. The dogs would fight the bear until the dogs or the bear were dead or the fight was stopped. People would bet on the winner. In England's Cruelty to Animals Act 1835, bullbaiting and bearbaiting were banned as inhumane. Enter dogfighting: a blood sport for gambling that was designed to fill the gap. As animal fighting—dogfighting and cockfighting—was neither a new sport nor one unique to England (it was already popular in Japan, the Philippines, and Mexico, to name a few examples), the bulldog, with its proven prowess at bullbaiting and bearbaiting, seemed like a good candidate to transition into the dogfighting arena. The bulldog was strong and loyal, had a strong jaw, and was an unrelenting attacker. But there was room for improvement.

For "improvement," the bulldog was crossbred with a terrier, a dog that is smaller and easier to handle than a bulldog and has an enormous prey drive. Their offspring, with its prey drive, size, strength, stamina, and love of people, which is important for the safety of human handlers, resulted in the perfect fighting dog and may have been the first custom-designed dog. The new breed was what we today call the American pit bull terrier. In this case, rather than crossbreeding to develop hypoallergenic qualities, the crossbreeding was intended to cause greater harm and promote a fight to the death, all for the entertainment of onlookers.

Because of the fear they invoked, these new fighting dogs appealed to the dogfighting world, as well as to criminals. I've seen many pit bulls with tough names like "Blade" or "Homicide" who've been used for such felonious purposes as guarding a crack stash or assisting in an armed robbery. This criminal element strove to breed the nastiest and scariest dogs on the planet. They did not try to preserve a love of people or encourage gentleness. The breeders mixed the meanest pits, and then mated their offspring with rottweilers to make them far larger and stronger than the breed's bulldog ancestors. There was no mindfulness applied to the breeding. The breeders weren't aware of the genetics of the dogs they mixed, and they weren't concerned about hurting parents or offspring.

Eventually the reputation of the American pit bull terrier, which, along with a few related breeds, is commonly called the pit bull, deteriorated to where it is today. Although many of them are actually wonderful pets, people are afraid of them, landlords often won't allow them, and many insurance companies refuse to cover them on homeowner policies. They're also frequently the subject of breed bans and other breed-specific legislation in cities or countrywide. In the United Kingdom, the pit bull terrier was one of four breeds banned in the country's Dangerous Dogs Act instituted in 1991. In the United States, there are bans on them in hundreds of cities across the country. There has, however, been a countermovement and twelve states have passed laws prohibiting pit bull bans.[2]

All dogs can bite, and a small dog can do a lot of damage, too (there was a case of a Pomeranian who killed an infant), but the mauling and fatalities caused by pit bulls and reported by the media have scared people into thinking that all pit bulls are sociopaths. The result is that they languish in shelters or are euthanized. The dog is paying the ultimate cost for the human intervention in its design.

There is a version of the pit bull bred today that is micro, but mighty. This small dog, usually custom-made at a buyer's request, resembles the pit bull, except that it is only about eleven inches tall, half the size of the average pit bull. The theory behind the making of this designer dog is that by shrinking the dog, you shrink the problem, i.e., you still get a pit bull, but it will be manageable. In reality, a micro pit bull will require the same amount of work as a larger pit bull, and is just as likely to be calm or aggressive. There's also no guarantee that making the dog smaller will allow an owner to be able to physically control it. You may save money on dog food, but the commitment, training, and caretaking necessary, as well as the medical bills, will not be considerably reduced.

The effects of the growing popularity of micro pit bulls are not yet clear, and their rise leaves me with many questions and concerns. Will they become increasingly desirable simply because they're a designer dog? For a person who has his mind set on getting a designer dog, will a micro pit bull be as desirable as a cockapoo? If more micro pit bulls are bred, will pit bull bans increase? And will they interest the nefarious people who breed

---

2   Dana Campbell, "Pit Bull Bans: The State of Breed–Specific Legislation," GPSOLO (July/August 2009), https://www.americanbar.org/newsletter/publications/ gp_solo_magazine_home/gp_solo_magazine_index/pitbull.html.

fighting dogs? Will these people see the micro pit bull and be inspired to create a ring of micro fighting dogs?

Shrink the Dog, Increase the Costs

Contrary to popular belief, shrinking a dog can actually increase its problems, as we can see in the now-raging craze for "teacup dogs." Teacups are typically miniatures of already little dogs like Chihuahuas, Pomeranians, and Yorkshire terriers. They're created by breeding the runts of litters together and continuing to breed the runts of the offspring until the resulting dogs are itty-bitty, teacup size.

Runts are often the unhealthiest of a litter and have medical issues associated with their stunted size; a breeder is exacerbating and amplifying their problems by bringing their negative traits forward to their offspring. The sound use of expert, science-based genetic protocols is virtually nonexistent in this breeding process and a breeder's conclusions regarding success are based solely on appearance, with no concern as to whether the dog behaves like a sociopath or suffers from a plethora of congenital or environment-related illnesses. And because legitimate breeders refuse to traffic in teacups, the supply is coming from questionable sources. Many teacups come from South Korean puppy mills.

Teacup breeders often seem to be competing against one another, each one out to make his dogs smaller. And they often have little interest in dealing with puppies who are not teacup in size, since the smaller the dog, the larger the price. When dogs intended to be teacups are born too large, they are often considered "unsatisfactory" and will quickly be sold if they can be. If not, they're ignored, abandoned, or destroyed.

The enormous medical problems teacups are often born with

are frequently exacerbated by a poor quality of life; for example, spending too much time in a purse and not enough time walking. I'm not sure who the first person to carry a teacup dog in her purse was, but celebrities are now regularly photographed doing this. It has become fashionable to do so. I'm only human, so I recognize and admit that teacups are adorable, like Tribbles and Furbys, but accessorizing with plush toys instead would be safer and more humane.

The attraction of teacups stems from what I call "perpetual baby syndrome," when people love kittens, but hate cats; puppies, but not dogs; and babies, but not teenagers. (The last one probably isn't just based on looks though!) Baby anythings are universally irresistible. On multiple occasions, spcaLA has encountered people who adopt kittens, turn them loose on the street when they're fully grown, and then return for more kittens.

So how do we stunt the growth of animals to keep them childlike? Breeding a teacup animal will do the trick, and teacup versions of a wide variety of animals are bred. There are dog and cat teacups, as well as pig and bunny teacups (often called dwarf pigs and dwarf bunnies), to name a few. Do an internet search and many more will come up. Opportunists around the world—puppy mill operators and other unscrupulous, unethical, and ignorant breeders—have figured out that there is high demand for these "forever infants," and a willingness to pay thousands of dollars for one.

Teacups tend to be treated like novelty items, objects, not living creatures. I often compare what has happened to them to what has happened to phones. Those of us of a certain age remember what it was like to have a house phone wired into the wall. We also remember the first mobile phones, which were about the size of telephone handles, about the weight of a brick, and stretched from your ear to your mouth. But from their humble beginnings, they morphed to smaller flip phones and then to small one-piece phones, about the size of a pack of cigarettes. If you recall the size of the phone that Ben Stiller used in *Zoolander*, about a square inch, you will understand what teacup breeders are trying to do. Watching the "need" for smaller and smaller phones makes me laugh, but there is nothing funny about living, breathing, emotion- and pain-feeling creatures being made smaller and smaller.

In case the teacup is not small enough, there are designer dogs who come in micro, mini, and pocket sizes (the aforementioned micro pit bull is one example). These tiny dogs are designed to impress, and the smaller the dog, the higher the price. TMZ reported that Paris Hilton paid $25,000 for a pair of tiny Pomeranians in January 2015, one reportedly weighing about six ounces and the other weighing about twelve ounces, and $13,000 for another mini Pomeranian in September 2015. In October 2016, she reportedly purchased an eight-ounce Chihuahua worth $8,000. The exorbitant prices and free red-carpet advertising make this an extremely profitable business for breeders. It also makes the dogs particularly prone to theft. Criminals kidnap (or dognap) them for resale, breeding, and ransom. Hilton's small Chihuahua Tinkerbell was dognapped in 2004 and she reportedly offered a $5,000 ransom.

Many owners of teacups and even smaller dogs may think there's no better display of elitism than a designer dog in a designer purse. But despite the costs and the early infatuation with these dogs, they are also prone to being abandoned. The dogs' health costs are expensive. And if a six-ounce dog pees and poops twice its weight in a $1,500 Louis Vuitton purse, the dog may transform from a best friend to a pest, from the best accessory to a defective one, and then he's shown the door.

CHAPTER 3

WHERE BABIES COME FROM: POORLY BRED VERSUS PURELY BRED

Horses are cheaper than oats.

—ANONYMOUS

Designer dogs are abandoned, turned into shelters, becoming sick, and dying at an alarming rate. The reason lies in the designer dog recipe.

Designer dogs can come from responsible breeders, but more often they're from puppy mills, backyard breeders (amateur breeders who often breed substandard dogs for fun or extra income), or hobbyists (who are like backyard breeders, but usually breed on a smaller scale and specialize in one breed). These suppliers exist all over the world and the dogs are sold domestically and internationally. The dogs sometimes travel twenty hours, halfway around the world, to get to their destination. Conditions are often squalid and infections spread between the puppies, further destabilizing their health and immune systems. It is a traumatic experience that can cause physical and mental health issues, or compound existing ones.

The internet offers quick and easy ways to find and buy designer dogs. The simplicity of this often leads to impulse-driven, whim acquisitions. And because the purchases are sight unseen, you can't be certain what you are actually buying and from whom. Often the dog you think you're buying, which you selected because of a cute photo or video, is not the dog that arrives at the door. Though existing laws protect these pets, there is frequently a lack of resources, officers, and the will to enforce them.

Some breeders and puppy mills are licensed; some are not. Some are registered with the United States Department of Agriculture (USDA); some are not. To be licensed and registered, a breeder needs to fill out a form, pay a fee, and agree to submit to inspections for compliance with minimum standards of care. Some breeders operate in places of business; others operate in homes. Proof of registration with the USDA or meeting an individual state's registration requirements doesn't guarantee compliance with federal, state, humane, husbandry, and retail laws.

After the 2016 presidential election, the USDA's Animal and Plant Health Inspection Service (APHIS) scrubbed much of the section of its website related to animal welfare. An "Editor's Note" on the website reads, "APHIS is implementing actions to remove documents it posts on APHIS' website involving the Horse Protection Act (HPA) and the Animal Welfare Act (AWA)."

On August 18, 2017, in response to a public outcry and threats of legal challenges, it reinstated the public search tool that provides access to AWA compliance records and some information regarding license and compliance issues, however, it's clear that the records are incomplete by the site's own admission that it will "continue to review records and determine which information is appropriate for reposting." Instead of transparently sharing all violation information known to it, the agency requires that the person seeking information file a request, stating, "Those seeking information from APHIS regarding inspection reports not currently posted to the website, regulatory correspondence, and enforcement-related matters may submit Freedom of Information Act requests for that information." This information is important for the public to be able to conduct even a minimal background check on the supplier of an animal being considered for purchase, and for those of us in law enforcement who use the information to inform state violations, assess patterns of mistreatment, and propose legislation. As of this writing, responses to information requests are slow, often ignored, and may be heavily redacted. New lawsuits have been filed to address these issues. The fact that this information is not readily available is reprehensible.

The Farmer in the Mill

After the Second World War, two things occurred that greatly affected the pet industry. The first was the post-war boom that ignited the rise of the middle class and the sudden presence of disposable income to spend on items like pets. The second was the changes in agriculture in the West and Midwest. New technology developed sophisticated agricultural machines and irrigation systems, and these, coupled with better pesticides, improved crops and crop yields, but there were also great challenges. Industrialization and urbanization caused many field hands to seek work in cities, and there were several years of severe drought during the 1950s. In response, cash-strapped farmers sought to produce new crops that would be immune to these challenges. The farmers began to "farm" puppies as an alternative cash crop. The farms became known as "puppy farms."

The farmers had no understanding of the dog-breeding industry and yet they jumped into business. To keep costs down, they stuffed dogs into wire chicken coops, crates, junked appliances, and car seats, and tied others to poles and tree trunks. To increase volume, wire cages were stacked high, one on top of the other, so that pee and poop seeped downward, to the cages beneath. The dogs bred and lived in their own waste and that of other dogs, and were subjected to extreme heat and cold. They were barely fed or medically tended to. Siblings were bred with siblings, parents, and any other dog that could be found. A single puppy farm could have hundreds or thousands of dogs imprisoned at any one time. The dogs were always ill, dirty, malnourished, and living in misery. Babies were often born deformed. Ones considered "unsuitable" (unlikely to sell or breed well) were quickly killed. These conditions continue today.

Capitalizing on post-war disposable income, dogs were offered for commercial sale. The first pet stores opened and puppies were put in store windows to lure customers. As is the case today, consumers wanted the breeds they saw in film, on television, and in the hands of matinee idols. For example, after the release of the film *Lassie Come Home* in 1943, the popularity of the collie increased by 60 percent during the next decade. The release of the movie *The Shaggy Dog*, in 1959, did the same for Old English sheepdogs, and this scenario played out countless times over the following decades.[3]

To satisfy demand, breeders quickly farmed and transported dogs. And dogs, legally property, became the new cash crop. Overhead costs were kept low in the "manufacturing process," and profits soared. It was a win-win for farmers and retailers, and consumers had easy access to whatever dog breed was in vogue.

In 1953, Patti Page recorded the song "How Much Is That

---

3   National Geographic, "Hot Dogs: America's Most Popular Breeds" (April 15, 2016), https://www.nationalgeographic.com/magazine/2016/americas-most-popular-dog-breeds/.

Doggie in the Window?" complete with barking sounds. The song, written by Bob Merrill and released by Mercury Records, hit Number 1 on both *Billboard* and *Cashbox* charts in 1953. Its single sold two million copies. Its chorus is unforgettable:

How much is that doggie in the window?
The one with the waggly tail
How much is that doggie in the window?
I do hope that doggie's for sale

This song celebrated the pure joy of a cute, loyal dog who protects his owner and provides companionship. It became the symbol of the premise that the pet store dog was the best thing in the world and that everyone should have one. Owning a dog from a pet shop was also thought of as a status symbol. In 1997, the song was adapted into a children's book, *How Much Is That Doggie in the Window?*, retold and illustrated by Iza Trapani, in which a little boy tries to save his entire allowance and chore money in order to purchase a "doggie in the window." Again, the iconic song encouraged and romanticized the purchase of a pet shop dog rather than the adoption of a shelter dog.

"Cheap to produce, expensive to buy" was the song the farmers were singing.

Over the years, the song "How Much Is That Doggie in the Window?" organically morphed into the anthem of activists who were decrying the dark side of puppy mills and who begged people to adopt from crowded shelters rather than buy from puppy mills. At issue for them was the inhumanity of the pet industry and the fact that healthy dogs were being euthanized for reasons related to time, space, and a lack of families to adopt them.

The anti–puppy-mill sentiment that became associated with

the song upset Patti Page so much that in 2009 she recorded a new version of the song titled "Do You See That Doggie in the Shelter?" Its lyrics were written by Chris Gantry and the rights for it were given to the Humane Society of the United States. The new version confronts the existence and practices of puppy mills and answers the question "how much" with "too much." In 2009, Patti Page was quoted by the Humane Society as saying:

> The original song asks the question: "How much is that doggie in the window?" Today, the answer is "too much." And I don't just mean the price tag on the puppies in pet stores. The real cost is in the suffering of the mother dogs back at the puppy mill. That's where most pet store puppies come from. And that kind of cruelty is too high a price to pay.[4]

The lyrics of "Do You See That Doggie in the Shelter?" were set to the same tune as the original version. It spoke of the dogs across the country without homes and owners to protect them, and who are lost and go hungry until rescued by a shelter.

Its catchy chorus is:

> Do you see that doggie in the shelter
> the one with the take me home eyes
> If you give him your love and attention
> he will be your best friend for life

While the public was celebrating its easy access to designer dogs, as the happy, tail-wagging music played on, back at the

---

4   The Humane Society of the United States, "Old Song Carries New Tune" (October 30, 2009), http://www.humanesociety.org/news/profile/2009/10/ old_song_new_tune_103009.html.

farms horrible things were happening. The original puppy farmers, who came from a background of agricultural crop farming and were not dog-breeding experts, treated the dogs like livestock rather than companion animals and figured out early on that the less they invested in good food, veterinary care, and husbandry, the higher their profits would be. They gave no regard to the health of mother dogs, which they bred constantly, or to genetic pool diversity. They also spent as little as possible to care for puppies during their transportation to pet stores. If some puppies died during transport, the price of the survivors would be raised so that the profit wouldn't be hurt.

The term "puppy mill," while colloquially used during the post-war period, wasn't formally introduced into legal vernacular until a 1984 Minnesota court case, *Avenson v. Zegart, 577 F. Supp. 958*. The plaintiffs in the case had been operating a dog-breeding business in Hubbard County, Minnesota, when, in March of 1982, they came to the attention of Lesley Zegart, then executive director of the Minnesota Humane Society, who was investigating dog-breeding businesses in Minnesota to determine which ones were operating as puppy mills. The term "puppy mills" got an official on-the-record definition in the courts' statement of facts, which defined it as "a dog breeding operation in which the health of the dogs is disregarded in order to maintain a low overhead and maximize profits." This characterization was true from the onset of puppy mills through the time of the 1984 case and continues to be true today.

The retailers the farmers shipped the dogs to were no better at caring for the dogs than they were. They too were out to make the highest profit margins possible, and they too crammed as many dogs into cages as they could and avoided paying for high-quality food or medical care. In effect they became minimills. If puppy deaths during transportation led to low inventory, that is, scarce "crop," or they had to pay more to farmers because of this, they, too, simply raised their prices to make up for it. Costs were passed to the consumer.

Many dogs died within days of arrival at a shop or purchase by a consumer. Yet stores were not legally obligated to provide refunds for the purchase cost or the medical bills incurred by grief-stricken consumers. Decades later, this persistent bad behavior inspired the passage of dog and cat "lemon laws" and "puppy/kitten mill warranty laws" in some states. These laws required refunds, reimbursement for medical bills, husbandry standards, and mandatory documentation of the source of each pet and its medical status—a customer bill of rights, if you will. Later in this book you will see actual cases involving these issues.

It is critical to note that although a puppy may survive early challenges and appear healthy, problems caused in the breeding process often manifest later in life, for example, medical disabilities, congenital problems, and incurable pain or discomfort. Sometimes an entire litter shares the same condition. Customers saddled with ongoing medical bills are often excluded from pet insurance protection policies, as many of the problems are considered preexisting conditions. The Affordable Care Act legislated against excluding coverage, or price gauging for coverage of people deemed to have preexisting conditions by insurance companies, but there is no such law governing pet insurance. Often when these costs are too burdensome, the pets are abandoned.

In the 1950s, stores like Sears, Roebuck and Company, and F. W. Woolworth Company started to get in on the action, featuring window dogs to attract customers, who bought other items as well once they were in the store. One could even order dogs from the Sears catalogue, foreshadowing the ability to order sight unseen from the internet.

What did department stores know about housing and caring for live animals? Not much more than anyone else at either end of the supply chain. I still remember the section of the Woolworth store that had what seemed like hundreds of little birds for sale crammed into cages!

As puppies were becoming available for purchase through a wide variety of retailers, brokers and dealers were developing a "convenience delivery program" to service the industry. The service would, if you will, "Uber" puppies to shops, buyers, and even research laboratories. It was a cash business with no regulation or third-party oversight. Puppies were transported in cramped quarters via pickup trucks or trailers and subjected to the elements.

In the 1950s, the beagle, made popular by the debut of Snoopy of *Peanuts* fame, became one of the most popular family dogs of the next two decades. Beagles also became, and still are, the favorite breed for animal testing in laboratories, as they are small, docile, and willing to please. The less stressed out a dog is during experiments, the fewer unpredictable changes in their physiology will occur and the more consistent the results will be. The calm demeanor of the beagle also reduces stress for the laboratory staff while they are experimenting on the dogs or caring for them. That a dog revered for his or her love and trust of humans is rewarded by being subjected to painful testing, seems like treachery and a cruel misuse of that trust.

In the late 1990s, the actress Kim Basinger brought cruel laboratory testing on beagles into the public eye when she learned that Yamanouchi USA, a pharmaceutical company, and Huntingdon Life Sciences, a medical research company, planned to break the legs of thirty-six beagles to test an osteoporosis drug.

Basinger mounted an enormous, well-publicized fight, during which she demanded the release of the dogs and even offered to take all thirty-six beagles home with her. The laboratory refused her request to remove the dogs, but it did stop the experiment.

To this day, the beagle remains the favorite dog of animal testing laboratories, but it's also the subject of "beagle freedom" legislation. The legislation calls for dogs (not just beagles) and cats to be released to humane societies and societies for the prevention of cruelty to animals, to be adopted into loving homes rather than euthanized, when they're no longer desired by the labs. In 2015, California enacted a law that required that publicly funded research institutions comply. There are comparable statutes in four other states: Connecticut, Minnesota, Illinois, and Nevada.

The bottom line is that the beagle suffered the torments of mass production because children wanted Snoopy dogs and medical researchers wanted to cut them up. Why do beagles still like us sixty years later? As an aside, it is not uncommon for me to see dogs who were brutally treated by their human companions still act like they love them. It is really heartbreaking to see that; we can't always understand that level of love and loyalty.

Pepper the Dalmation

With the surge of family pets came a surge of burglaries. Family pets were easy targets because they are more docile than stray dogs. This quality made them especially desirable for research institutions, since it's easier to experiment on a dog with a calm demeanor. And so, dognappers caught pet dogs and sold them to research facilities.

In 1965, the issues of stolen family pets ending up in laboratories, their mistreatment there, and the deplorable treatment many of them received early on from puppy farm operators and other breeders, became part of the public discourse after the journalist Coles Phinizy published a story titled "The Lost Pets that Stray to the Labs" in the November 29, 1965 edition of *Sports Illustrated*.[5] It told the tale of Pepper the Dalmatian.

Pepper was a five-year-old dalmatian and the beloved pet of Julia and Peter Lakavage and their children when she disappeared from their Pennsylvania farm. The couple advertised that Pepper was missing and searched relentlessly for her. Ultimately, they learned of the arrest of William Miller, an opportunist pet dealer, who was attempting to ship dogs and goats in a truck that was not up to local vehicular codes. His shipment was housed for a night at a Pennsylvania SPCA while he resolved his truck violations. In the account of the arrest, the Lakavages read that there were two dalmatians in the truck. Julia called the shelter to ask about the dalmatians and later, after seeing a photo of them, identified one as Pepper. But before the Lakavages could get to the shelter to rescue their dog, Miller had already unloaded his shipment at Montefiore Hospital in New York City. When Julia learned where Miller had gone, she raced to the hospital, but by the time she got there, it was too late. Pepper had died during experimentation and been cremated after. It was too late to save her.

During an investigation by Pennsylvania police officers, Miller claimed to have received the dog from a person who claimed he got the dog from another person, and so on. There were no reliable records to document any transactions. It was clear that dealers were stealing, selling, and transporting dogs across the country, unregulated.

---

5   Coles Phinizy, "The Lost Pets that Stray to the Labs," *Sports Illustrated* (November 29, 1965), https://www.si.com/vault/1965/11/29/612645/ the-lost-pets-that-stray-to-the-labs.

During the hunt for Pepper, Senator Joseph Clark of Pennsylvania and Congressman Joseph Resnick of New York learned of the theft and investigation. With Senator Clark's support, Congressman Resnick introduced a bill, H.R. 9743, that mandated that anyone dealing, buying, selling, or brokering dogs be licensed by the federal government and keep a proper record of all related transactions.

The proposed bill led to a raging debate in Congress between two sides: people who had witnessed extreme animal cruelty in puppy farms or those like the Lakavages, who suffered heartbreaking experiences, against members of the medical research community who wanted test subjects regardless of the source, whether bred for research, stolen, or bought from public pounds or private shelters. (The terms "pound" and "shelter" are distinctions without a difference. Antiquated statutes still in use tend to use the word "pound" while modern ones have changed the term to "shelters." Sometimes the former connotes a public rather than private organization.) Some legislators held a strong position for federal action and others advocated states' rights and said the federal government shouldn't intervene. Phinizy articulated the core of the conflict when he wrote:

> Whether or not the martyred Pepper will succeed in making a federal case out of dognapping is up to the men who make our nation's laws, but there are two things that the legislative investigation of her death and disappearance have made quite clear: 1) many pet dogs are being stolen from the front lawns and sidewalks of this country, and 2) the thefts in large part are motivated by science's constant and growing need for laboratory animals.[6]

6   Ibid.

The bill ultimately passed, the winning argument being that because dogs were shipped throughout the country, consistency in regulation was necessary and so the legislation must come from the federal government and not the states. But it wasn't an easy win.

In his article, Phinizy lamented what seemed like a catch-22 presented to Congress, that they had an important subject to regulate, but insufficient statistics to guide their decision. In other words, there was no documentation to prove the need or lack of need for documentation, and no documentation to prove the activity at all. He wrote:

> At a preliminary hearing on the bill some weeks ago a great many charges were made, but not much was proved one way or the other. The truth of the matter is that the whole business of dog procurement for laboratory use, illicit or otherwise, wallows in a sea of insufficient fact. How many dogs do US laboratories use in a year? Nobody knows. How many laboratories use dogs in experiments? Again nobody knows. Where do most of the dogs come from? No one can say for sure.[7]

The problem articulated by Phinizy is one that the animal welfare community sees over and over again. We learn of atrocities being committed regularly and we demand change to prevent the atrocities from recurring, changes like requiring all parties to formally document their activities so there is a paper trail, and responsibility can be allocated if something goes wrong, but because of the lack of authentic, accurate, and uniform documentation of the sketchy transactions of breeders and dealers, it's hard to prove their activities, and there is no documentation

---

7   Ibid.

to rely on to support the enactment of laws that would mandate documentation. The lack of statistics makes it impossible to gauge the size of a problem, if any, and the extent of the necessary fix if one is needed.

The right jab leveled by the *Sports Illustrated* story against the pet breeding and dealing business was followed by a left cross when, on February 4, 1966, *Life* magazine published "Concentration Camp for Dogs," a photographic essay by Stan Wayman that depicted heart-wrenching images of the cruel confines of a puppy farm.[8] Wayman had taken the photos during a raid by the Baltimore Humane Society and Maryland State Police on a compound holding 103 dogs. One page of photographs was titled "Raiders Discover a Den of Woes." One of the many upsetting images showed a dog who had frozen to death in below freezing temperatures. The public outrage from these two seminal articles was loud and clear enough to be heard by the legislators on Capitol Hill and President Lyndon B. Johnson. Sadly, those haunting photographs could have been taken yesterday, as puppy mills like the one photographed still exist around the country.

On August 24, 1966, President Johnson signed the Laboratory Animal Welfare Act of 1966 (now known as the Animal Welfare Act (AWA)), which incorporated the bill proposed by Clark and Resnick, and added humane treatment standards for laboratory animals. Upon signing the bill, President Johnson remarked:

> Progress, particularly in science and medicine, does require the use of animals for research and this bill does not interfere with that. But science and research do not compel us to tolerate the kind of inhumanity which has been involved in the business of

---

8   Stan Wayman, "Concentration Camps for Dogs" (February 4, 1966), *Life*.

supplying stolen animals to laboratories or which is sometimes involved in the careless and callous handling of animals in some of our laboratories.[9]

The AWA provided guidelines for minimum standards of care for the humane treatment of animals in housing, transport, and breeding, as well as a slew of documentation and record-keeping requirements with detailed transaction records and source, veterinary, and sale records. It also outlined two types of parties, breeders and dealers, and required licenses for them, a class A license and a class B license, respectively.

The definition of a class A licensee includes the following language:

> Class "A" licensee means a person subject to the licensing requirements . . . and meeting the definition of a "dealer" [a person breeding, buying, and selling] and whose business involving animals consists only of animals that are bred and raised on the premises in a closed or stable colony and those animals acquired for the sole purpose of maintaining or enhancing the breeding colony.[10]

This type of licensee can sell only what he breeds on his premises. It doesn't mean the licensee doesn't run a puppy mill, and is breaking the guidelines for minimum standards of care, but theoretically the source of the animals and the breeding records

---

9   "Lyndon B. Johnson, 402 – Remarks Upon Signing the Animal Welfare Bill" (August 24, 1966), http://www.presidency.ucsb.edu/ws/index.php?pid=27796.

10 United States Department of Agriculture, National Agricultural Library, "Final Rules: Animal Welfare; 9 CFR Parts 1, 2, and 3," Federal Register, Vol. 54, No. 168 (August 31, 1989), https://www.nal.usda.gov/awic/final-rules-animal-welfare-9-cfr-parts-1-2-and-3.

are established and documented, so a license might be secured. A class A licensee could be breeding specifically to sell either to a laboratory or the pet industry.

Later in the act, the language states that a license is not required for breeders who have four or fewer breeding females and sell only the offspring of these females.

The description of a class B licensee includes the following language:

> Class "B" licensee means a person subject to the licensing requirements . . . and meeting the definition of a "dealer" and whose business includes the purchase and/or resale of any animal. This term includes brokers, and operators of an auction sale, as such individuals negotiate or arrange for the purchase, sale, or transport of animals in commerce. Such individuals do not usually take actual physical possession or control of the animals and do not usually hold animals in any facilities. A class "B" licensee may also exhibit animals as a minor part of the business.

This licensee is a random source dealer, which means his animals could come from anywhere, including front lawns, pounds, owners, swap meets, and animal shelters. Some of these dealers work with middlemen called "bunchers." A buncher is not licensed, permitted, or otherwise regulated and can pick up pets from regulated or unregulated sources, like Craigslist or "free to good home" advertisements. The buncher sells to a class B licensee, who can then sell to anyone, including research institutions. The buncher rarely has documentation and is not required to. Hence, a class B licensee, though regulated, can circumvent requirements and muddy the footprints on the source trail by using a buncher.

A class B licensee, though he or she usually doesn't hold animals at a facility, may store the dogs in the back of a truck or use some makeshift confinement system until enough animals are acquired to move to the research institutions. The conditions under which these animals are held are frequently grossly inadequate and the dogs often become ill because of this, or it compounds existing illnesses. The standard defense is to assert that the dogs were in poor shape when they were acquired, so the dealer should not be held responsible. Since the dogs could come from such places as auctions and pounds, it is hard to prove this defense false. Also, no one really asks. It's the silent wink and nod of the seedy underbelly of the business.

Consider this, if Pepper the Dalmatian was stolen by a buncher and sold to a class B licensee, a random source dealer, who then sold Pepper to a laboratory, the result at that time would have been the same.

The AWA was flawed, and enforcement lax, but it was a start. And there were ripple effects as some states enacted, revised, or augmented their own animal welfare laws. At the time, in the mid- to late 1960s, there was some optimism that relief for these dogs would come. Sadly, it was not to be.

"Try It On the Dog"—Pound Seizure

The AWA was not seriously enforced by the USDA or law enforcement, and there was little respite for dogs. They continued to be brutally farmed, stolen, sold, and abused. Pound seizure practices exacerbated the harm to them.

Pound seizure is the sale or release of pets from a pound to a research testing or educational facility. In this way, shelters, ostensibly safe havens for animals, betrayed them as well. The shelters were supposed to be places where lost pets could be kept safe until found, and abandoned animals could be cared for. Shelters were supposed to help heal dogs who'd been injured or abused, and, ideally, place them into new and loving homes. They were also places that could provide a peaceful, painless, and humane death to animals should that be necessary. Their overarching goals were to ease suffering, provide shelter from the elements, and protect dogs from harm. Or at least they proclaimed these were their goals.

The biomedical and pharmaceutical industries still wanted to test on animals, and researchers and laboratories discovered that pounds were a cheap source of dogs. For their experiments, they did not require that dogs be purebred, or even bred in a consistent manner, to achieve uniform traits. They only required that they be dead or alive.

In the 1940s, several laws were enacted that mandated that public pounds (and even some private organizations housing animals) turn over unclaimed animals to research institutions for experimentation and testing. The following is an actual law passed in Minnesota in 1949 and is representative of similar laws passed in other states around the same time.

35.71 UNCLAIMED AND UNREDEEMED ANIMALS IMPOUNDED; SCIENTIFIC USE.

Subdivision 1. Institution defined. As used in this section, "institution" means any school or college of agriculture, veterinary medicine, medicine, pharmacy, dentistry, or other educational or scientific establishment properly concerned with the investigation of, or instruction concerning the structure or functions of living organisms, the cause, prevention, control or cure of diseases or abnormal conditions of human beings or animals.

Subd. 2. Application by institution for license. Such institutions may apply to the State Live Stock-Sanitary Board for a license to obtain animals from establishments maintained by or for municipalities for the impounding, care and disposal of animals seized by lawful authority. If, after investigation, the State Live Stock Sanitary Board finds that the institution making request for licensure is a fit and proper agency within the meaning of this section, to receive a license, and that the public interest will be served thereby, it may issue a license to such institution authorizing it to obtain animals hereunder, subject to the restrictions and limitations herein provided.

Subd. 3. Supervisor of licensed institution. It shall be the duty of the supervisor of any establishment referred to in subdivision 2 to make available to an institution licensed hereunder, from the available impounded animals seized by lawful authority, such number of animals as the institution may request, provided however, that such animals shall have been impounded for not less than five days or for such other minimum period of time as may be specified by municipal ordinance and remain unclaimed and unredeemed by their owners or by any other person entitled to do so. If a request is made by a licensed institution to such supervisor for a larger number of animals than are available at the time of such request, the supervisor of such establishment shall withhold thereafter from destruction, all such unclaimed and unredeemed animals until such request has been filled, provided that the actual expense of holding such animals beyond the time of notice to such institution of their availability, shall be borne by the institution receiving them.[11]

Other pound seizure laws such as the one above were enacted before the Animal Welfare Act, so there were no restrictions or rules concerning anesthetizing or treating animals humanely while subjecting them to experimentation.

The practice of pound seizures allowed shelters to monetize their animals in two ways. A shelter could sell a live animal to a research institution rather than pay to house and care for it until an owner claimed it as a lost pet, or an abandoned animal found a new home. Or a shelter could euthanize a pet sooner rather than later, as dead pets could also be sold to institutions. When word got out about these practices, some members of the public who knew and disapproved of them would release or leave strays to suffer and die on the streets rather than turn them into pounds. This fundamentally gutted the primary objective of shelters and deprived the animals of any solace.

It was not until 1993, almost fifty years after pound seizure laws were passed and more than thirty years after the label of a class B dealer was established, that a five-day holding period was enacted, as an amendment to the Animal Welfare Act, ostensibly to allow families to reclaim their lost or stolen pets before it was too late. During this period, laboratories (who sold to other institutions such as laboratories and schools), institutions, and shelters were prohibited from selling pets to a dealer. Such a provision might have saved Pepper the Dalmatian thirty years earlier. The following is the USDA summary statement:

---

11 The Office of the Reviser of Statutes, "35.71 Unclaimed and Unredeemed Animals Impounded; Scientific Use or Other Disposition," 2011 Minnesota Statutes, https://mn.gov/boards/assets/BAH%20Rule%20Book_tcm21-26580.pdf.

SUMMARY: We are amending the regulations under the Animal Welfare Act (Act) to require that dogs and cats acquired by pounds and shelters owned and operated by States, counties, and cities, private entities established for the purpose of caring for animals, such as humane societies or contract pounds or shelters, and research facilities licensed as dealers by the United States Department of Agriculture, be held and cared for at those establishments for at least 5 days before being provided to a dealer. We are also amending the regulations to require that dealers provide a valid certification to anyone acquiring random source dogs and cats from them. These amendments are being made pursuant to the most recent amendment of the Act. The amendment to the Act was enacted to prevent the use of stolen pets in research and to provide owners the opportunity to locate their animals.

EFFECTIVE DATE: August 23, 1993.[12]

As of this writing, lest you think that pound seizure is a dead issue, only eighteen states and Washington, DC, have prohibited the practice. Thirty-one states leave the decision to local cities and counties, and Oklahoma still requires that animals be handed over for testing. California, considered the most enlightened and progressive state in issues of animal welfare, did not ban the practice of turning over live animals to research institutions until 2016. However, California shelters may still turn over dead animals for research, if they post a notice informing the public that they do. (The notice requirement used to apply to live animals

---

12 United States Department of Agriculture, National Agricultural Library, "Final Rule: Random Source Dogs and Cats," Federal Register, Vol. 58, No. 139 (July 22, 1993), https://www.nal.usda.gov/awic/final-rule-random-source-dogs-and-cats.

as well.) The State Humane Association of California, now the California Animal Welfare Association, of which I am currently president, sponsored a bill to ban turning over animals, both dead and alive, to research facilities (Assembly Bill 2269). The part of the bill protecting live animals became law, but the part intended to protect dead pets and let them rest in peace did not prevail.

If you want to see where your state stands on the issue, visit the website of the American Anti-Vivisection Society, aavs.org, which summarizes current pound seizure laws in each state.

In 1970, the congressional statement of policy regarding the AWA and its message to the USDA changed. When the act was enacted in 1966, the congressional statement of policy read:

> Progress, particularly in science and medicine, does require the use of animals for research and this bill does not interfere with that. But science and research do not compel us to tolerate the kind of inhumanity which has been involved in the business of supplying stolen animals to laboratories or which is sometimes involved in the careless and callous handling of animals in some of our laboratories.[13]

Enacted under President Richard Nixon, the 1970 amendment, which is still in place today, reads:

> The Congress finds that animals and activities which are regulated under this chapter are either in interstate or foreign commerce or substantially affect such commerce or the free flow thereof, and that regulation of animals and activities as provided in this chapter is necessary to prevent and eliminate

---

13 The Lyndon Baines Johnson Presidential Library, "On This Day in History: August," http://www.lbjlibrary.net/collections/on-this-day-in-history/august.html.

burdens upon such commerce and to effectively regulate such commerce, in order—

(1) to insure that animals intended for use in research facilities or for exhibition purposes or for use as pets are provided humane care and treatment;

(2) to assure the humane treatment of animals during transportation in commerce; and

(3) to protect the owners of animals from the theft of their animals by preventing the sale or use of animals which have been stolen.

The Congress further finds that it is essential to regulate, as provided in this chapter, the transportation, purchase, sale, housing, care, handling, and treatment of animals by carriers or by persons or organizations engaged in using them for research or experimental purposes or for exhibition purposes or holding them for sale as pets or for any such purpose or use.[14]

The original language stressed the inhumanity of the business of supplying animals for research and said that the need for such research does not compel us to tolerate suffering. The focus was on stopping the inhumanity, whether it was suffered by the animals or by those who lost their pets to random source dealers. There was an attempt made to balance the need for medical research and the reluctance to sacrifice animals for it. The

---

14 U.S. Department of Health and Human Services, The Office of Research Integrity, "ORI Introduction to RCR: Chapter 4. The Welfare of Laboratory Animals," 7 USC, 2131-2156: Animal Welfare Act as Amended. https://www.law.cornell.edu/uscode/text/7/2131.

balancing act was to recognize that medical research is necessary, that science requires animal testing, but that animals must still be treated humanely.

The amended language, just four years later, shifted the priority of the policy so that it became a way to prevent and eliminate the burden on interstate and foreign commerce. The policy still called for the humane treatment of animals, but its new message was loud and clear: Regulations to treat animals humanely may occur as long as a free flow of animals is maintained and no burdens on interstate and foreign commerce that would retard this flow are created as a result. The balancing act shifted to balance regulation against the need for the free flow of commerce.

Understanding this helps shed light on the reluctance to enforce laws that stymie efforts to move "product" and discourage the licensees from working; it also answers the question of why there is no change in practice since the death of Pepper the Dalmatian. The emphasis on preventing a burden on commerce also informs the continued practice of pound seizure, the continued purchase of animals from class B random source dealers, and the lack of mandatory upgrades in the standards of care and husbandry of puppies and their mothers. Complying with such regulations would cost money, reduce profits, and impair the ability to move high volumes of animals quickly.

Sadly, not a lot has changed since 1966. Conditions at puppy mills are just as gruesome. Class B dealers and pet thieves still operate relatively unchecked, and puppies continue to suffer.

In July 2000, the *Atlantic* published an article by Judith Reitman, titled "From the Leash to the Laboratory: Medical-research institutions draw on a thriving black market in stolen and fraudulently obtained pets." In the article, Reitman shared her experience in Poplar Bluff, Missouri, on "trading day," a weekly event when dogs are sold and traded to middlemen who then deliver them to buyers, including but not limited to those representing medical research institutions. (Other purchasers might buy for such things as puppy mills and organized dogfights.) The article revealed how the story of the puppy business was the same as it was in the 1950s, including the ill health of the dogs being sold, despite the enactment of the Animal Welfare Act, the rise in public condemnation of the high-volume breeding business, the enactment of stricter state legislation, and enforcement tools. The type of sad events Reitman described continue to this day. She wrote of vehicles from assorted states laden with crates, each crammed with purebred and mixed-breed dogs in tiny cages, and described other disturbing scenes:

> Men sweating under feed caps were pulling dogs out by their legs or muzzles. Many of the dogs were emaciated, their bellies swollen from worms or other parasites, their coats matted with their own feces and urine. The scene was hauntingly quiet. When a dog did bark, it was reproached with a swift kick.

> Around eight-thirty a nondescript white van pulled onto the lot. The driver—a registered dog dealer, licensed by the US Department of Agriculture—swung open the loading doors, revealing dozens of empty metal cages. About a hundred dogs were for sale on the lot. Soon sellers were clustered around the dealer's van. The day's trading had begun.[15]

Reitman paints a scene where no one seems concerned with the condition of the animals and cites a source that flat-out said

---

15 Judith Reitman, "From the Leash to the Laboratory," The Atlantic (July 2000), https://www.theatlantic.com/magazine/archive/2000/07/from-the-leash-to-the-laboratory/378269/.

that the USDA considers the licensees their constituency, which creates a conflict for them: should they bust their own constituency?

Reitman's characterization of the scene as "hauntingly quiet" was a significant observation. Many of us who have seen dogs in these situations have also remarked on the eerie silence of the dogs. Any dog owner hearing this quiet would recognize the eeriness of it. Dogs bark at everything, especially other dogs, and small-breed dogs and puppies will yap endlessly at their own shadows, so the lack of noise is of evidentiary value in an animal cruelty case. A dog's silence can be a symptom of mental anguish, depression, and surrender. Suffering, an element of animal cruelty statutes in all fifty states, encompasses psychological as well as physical pain. I know from firsthand experience that once we take dogs out of a puppy mill and care for them properly, they perk up and begin to display the normal annoying behavior that we know and love. They bark and communicate like dogs should.

Reitman ended her article with a moving description of the close of trading day that conjures up a bleak and desolate image of a dystopian slice of life.

> By ten o'clock the back lot was deserted—a patch of dust under a flat sky. Several boys were combing the ground for coins and cigarette butts. But there was little to salvage: just a couple of dog collars attached to a metal chain.

The way she captured this scene is brilliant. What message is being conveyed to the young children who are watching and learning from what they witnessed that day? How desensitized to suffering have they become? Do they think that emaciated dogs with swollen bellies are normal and that all dogs look that way?

Where do they learn to empathize with other living creatures? Perhaps one of the reasons why the future keeps repeating the past is that we are passing past values forward, from generation to generation. This is where designer dogs come from.

CHAPTER 4

THE PERFECT STORM: IT'S POURING DESIGNER CATS AND DOGS

It ain't what a man don't know that makes him a fool, but what he does know that ain't so.

—Josh Billings

To understand the perfect storm that would shake up the puppy market industry, create a pet trafficking nightmare, and continue the atrocious treatment of puppies in every aspect of this business, it is critical to recognize a series of converging events: the advent of designer dog mania, the ease of selling across borders, both interstate and internationally, and the ability to instantly purchase a puppy from anywhere in the world via the internet.

Not long after the change in congressional policy to promote free trade, in the late 1980s, the labradoodle was invented and the desire to have a custom-made dog in all its iterations took off, and with this came heavy manufacturing to meet the demand. By 2010, the craze was in full swing, and one of these designer mutts, the goldendoodle, was the most popular and coveted dog that year, followed by the labradoodle and the puggle, a cross between a pug and a beagle. Once the expensive pets became family members, it became common for them to travel with families on vacations. The United Kingdom relaxed its border restrictions in 2012 to facilitate travel with pets; that opened the flood gates for pet trafficking into and out of the country. Pets could just as easily be taken across its border for insidious reasons as they could for kinder ones.

On October 6, 2010, Instagram made its debut and presented an entirely new form of social media, one based entirely around visuals. Facebook and Twitter were the most popular social media platforms at the time, but they were primarily text based. Soon people were posting all forms of photographs, most of them centered on their lives and personal experiences. Then came celebrity photos and, of course, celebrity dog photos. With just a few point-and-clicks anyone could see the photos and then locate a place to order an imitation. They could also request a custom-made creature, like an "allergy-free" dog or a dog they thought might be a hit on Instagram, teacup dogs for example.

To satisfy demand, breeders mixed every breed with poodles, and teacup breeders competed to see who could breed the teeniest dog, as if they were novelty items. And everyone with a god complex set out to create a new breed. People were willing to pay exorbitant prices for their very own designer dog, particularly if they thought it would be an Instagram sensation. Celebrities provided free advertising and endorsements of designer dogs, and paparazzi shots of celebrity and designer dog duos accentuated the appeal.

The speed of buying on the internet was light-years beyond ordering from the Sears catalogue. Times had changed, and yet the business model of the 1940s puppy farm was still in use, if not made worse, as puppy mills materialized all over the world to satisfy the demand and participate in get-rich-quick schemes. Puppy mill owners still bred cheap and sold high, and dogs were still stolen and resold—still to laboratories, but now also to a vibrant black market of wannabe owners looking to pay less than full retail value.

I should mention that while dogs were and continue to be the core of the designer pet industry, a variety of other animals continue to be produced in mills. Kitten mills, for example, are also heartbreaking places of mistreatment and cruel genetic experiments. The twisty cat or "squitten" is a cat bred specifically for the deformity of stunted forelegs, causing, in theory, the cat to look like a cross between a cat and a squirrel. The "lykoi" is a so-called werewolf cat that looks scary and acts like a dog. Breeding them is another genetic mutation game with sad and painful results.

Today, as it continues to pour designer cats and dogs, the USDA, with its lack of oversight, does not have accurate information about the industry. The USDA website lists 5,863 facilities licensed to operate, which includes places such as zoos, auctions, sanctuaries, and training facilities. The website states that there are 1,853 licensed facilities that include research institutions and other establishments. The numbers are not accurate or even close to the existing number of puppy mills, breeders, dealers, and sellers, as the USDA inventory does not include those who don't bother to register, the opportunistic backyard breeders and hobbyists, the puppy mills worldwide that ship puppies to buyers in the United States, or internet pet shops that sell directly to anyone with a credit card. No records are kept of these.

It follows that the USDA doesn't provide any meaningful enforcement. They disregard facilities known to be functioning without a license and don't hunt for others. The USDA receives graphic complaints that suggest that there has been no meaningful improvement since 1966 in the conditions of puppy mills, and they hear people's frustrations with the trafficking of puppies into and within the United States. In response to this, they created some new regulations, effective in 2014, and in some areas have begun collaborating with other law enforcement agencies. But despite the flowery language on their website about treating animals humanely, their practice is still primarily concerned with providing support, in the way of free, unencumbered commerce, to their registered licensees and revenue generators, and allowing the free flow of commerce to continue unencumbered.

Other revenue generators, like taxpayers, support the humane treatment of animals, and are responsible for far more revenue income than the piddling amount from license fees, yet that doesn't seem to matter.

I had the honor of appearing on the *Daily Show* in 2010, when a bill to improve puppy mills was on the Missouri ballot. A spokesperson opposing the law was asked why dogs were still stacked inhumanely on top of one another in wire cages. The response was, "We stack humans in apartments all the time, in all the cities." The audience gasped. The spokesperson compared the bill to Obamacare, saying, "They're expecting all the breeders to sit there and pay for exorbitant amounts of care that are not needed, like adequate food, adequate water, adequate space." Again the audience was shocked. The bill was passed, but later overturned by the Missouri State Senate.

In 2013, the USDA did take a step in the right direction when it sought to regulate the sale of pets over the internet. Until 2013, this sort of transaction fell between the cracks and was essentially invisible to law enforcement. In theory, puppy mills and dealers need to be registered and licensed by the USDA and are subject to inspections, while brick-and-mortar retail pet stores are subject to regulation and inspections in their communities under local law.

Therefore, someone should have eyes on the animals at one end or the other, if not both, as many pet stores order their inventory from puppy mills. However, ordering from someone not licensed, or delivering directly to a consumer's home, avoids scrutiny at both ends of the transaction and, as such, went unregulated. Naturally, sick, injured, and genetically inferior animals were being sold to naive buyers across the country with no checks and balances, and not held to any minimum care standards. This situation was not good for the consumer and was far worse for the animals.

To address the problem, the USDA revised the forty-year-old Animal Welfare Act's definition of "retail store" to require that the seller and the buyer be in the same place at the same time with the actual animal being sold. A video or photo does not meet this requirement. It also added language that gave businesses that sell pets over the internet (sight unseen) their own category and made them subject to licensing and inspection under the Animal Welfare Act. In making the changes, the USDA restored the definition of "retail pet store" to its original intent, which is "a place of business or residence at which the seller, buyer, and animal available for sale are physically present so that the buyer may personally observe the animal and help ensure its health prior to purchasing or taking custody of it."[16]

The USDA website explains the change in this way:

With the growth of the Internet in the 1990s, technology brought with it new and unforeseen opportunities to buy and sell pets. More retailers began offering pets for sale sight unseen

---

16 United States Department of Agriculture, Animal and Plant Health Inspection Service, "USDA Restores Important Check and Balance on Retail Pet Sales to Ensure Health, Humane Treatment" (August 29, 2016), https://www.aphis.usda.gov/aphis/newsroom/news/sa_by_date/sa_2013/sa_09/ct_retail_pet_final_rule.

and to sell and ship them nationwide. While pet animals were sometimes sold sight unseen via telephone and mail order decades before passage of the AWA, the Internet has made it possible for many more persons throughout the United States to buy pets online from retailers without ever having to be physically present at the seller's place of business or residence and personally observe the animals offered for sale as the AWA intended. With the dramatic rise in sight unseen sales have come increasing complaints from the public about the lack of monitoring and oversight of the health and humane treatment of those animals.

In order to ensure that the definition of retail pet store in the AWA regulations is consistent with the AWA and that all animals sold at retail for use as pets are monitored for their health and humane treatment, we published . . . a proposal to revise the definition of *retail pet store* and related regulations to bring more pet animals sold at retail under the protection of the AWA.[17]

Today, sight unseen vendors have to be licensed and inspected by the USDA's Animal and Plant Health Inspection Service to ensure that the pets they sell to the public receive minimum standards of care and that other requirements prescribed by the AWA are followed. There are a host of exemptions and other changes in this set of revisions, but the requirements were a step in the right direction. It was recognition of the overwhelming number of sick animals from larger-scale breeders that were delivered to consumers' homes with absolutely no oversight, but plenty of heartbreak. Sadly, as previously mentioned, the USDA reversed

17 Ibid.

the progress in 2017 when it scrubbed its website of compliance and complaint data. Since then, it's been more difficult for consumers to check if a seller has been charged with violations before making a purchase, and it's been harder for pet shops to comply with laws that mandate such checking. If a person files a Freedom of Information Act request, the puppy in question may be an adult by the time the information is obtained, if it is at all. By making compliance impossible, this new policy amounts to a de facto repeal of the laws protecting consumers.

In several states, it is illegal for dealers or puppy mills to sell puppies to pet stores unless they are registered with the USDA and the pet shop has documented that it checked for violations. Some states mandate that pet stores only buy from class A breeders with no violations. Former Virginia governor Terry McAuliffe signed a law that took effect July 1, 2017 that requires that every broker and seller who provides dogs to Virginia pet shops have a valid USDA license, no citations for a critical violation, and no more than three noncritical violations in the two years prior to receiving a dog. These are all steps in the right direction, but if it is up to pet stores to demand violation information from dealers, then the system being relied on is basically an honor system. With an honor system and a world of people whose only interest is making money, what could go wrong?

CHAPTER 5

HOW MUCH IS THE LEMON IN THE WINDOW?

Teaching kids to count is fine, but teaching them what counts is best.

—BOB TALBERT

The December 3, 1989 issue of the *Los Angeles Times* featured an article by Nicole Yorkin titled, "The Pet-Store Controversy— How Sick Is That Doggy in the Window?"[18] In it, Yorkin tells the story of Margaret and Al Valdez, a couple who bought a $600 miniature pinscher puppy in a pet shop at the local mall. The puppy was brought home and named Rambo, and then immediately commenced playing with their three young sons. Soon after, Rambo began vomiting blood. Forty-eight hours later, he was dead. The cause of death was canine parvovirus. The pet store denied the family a refund, but offered them another miniature pinscher, who became Rambo II. Rambo II was missing some hair along his backside, which the store assured the family was normal.

When the family took Rambo II to the vet, the dog was diagnosed with mange and ringworm. They took him back to the

---

18 Nicole Yorkin, The Pet-Store Controversy - How Sick Is That Doggy in the Window?"
    Los Angeles Times (December 3, 1989), http://articles.latimes.com/1989-12-03/magazine/
    tm-0_1_pet-shops.

pet store and were told to leave the dog, and that they would get a call to retrieve him after the store's vet declared him healthy. When they went to retrieve the dog two weeks later, he was almost completely hairless. The family declined to keep Rambo II. Again, the store would only offer an exchange, not provide a refund.

A few weeks later, the store told the family that another miniature pinscher had arrived and offered them the puppy. The store had the gall to ask for an additional $100, as this dog came from a different breeder. The couple was furious, but ultimately made the purchase, since their children were distraught and traumatized by having loved and lost two pets in a short amount of time. Rambo III also had mange and ringworms, and he had a heart murmur. The family kept Rambo III, and got sacked with heavy veterinary bills. Their experience, buying what looks like a cute, fun-loving dog, and then finding out that he is very sick, is common. I've heard and seen similar accounts over and again throughout my career in animal welfare.

The Lady and the Tramp Case

In 1996, California enacted a "pet protection/puppy lemon law" and a "puppy mill warranty law." Officially named, respectively, the Lockyer-Polanco-Farr Pet Protection Act and the Polanco-Lockyer Pet Breeder Warranty Act, the laws were a response to the horror of puppy mills, recognition that most pets came from one, and that there was a lack of consumer protection when purchasing a pet. The laws sought to address these issues by mandating humane care and husbandry standards for pets and providing consumers with rights and recourses should they be sold a physically or mentally ill pet, a lemon, from either a store or a local breeder, puppy mill or not. The legislation said that violators could be held accountable both civilly and criminally.

spcaLA prosecuted the first case under this legislation. The criminal investigations that led to the case began in November 1995, after spcaLA's elite enforcement unit received multiple complaints from people who had purchased expensive dogs (ranging between $700 to $2,500 depending on the breed) who turned out to be blind, deaf, or gravely ill. Some of the dogs died shortly after purchase. When the buyers complained, they were ignored by the sellers. For spcaLA's investigation of the complaints, its humane officers (officers with all the normal powers of peace officers, including making arrests, serving warrants, and carrying weapons) closed in on a family which bred dogs and also acted as an intermediary between other breeders and potential customers. The officers found cocker spaniels, like the dog Lady from *Lady and the Tramp* (Tramp was a junkyard mutt), and other small-breed purebreds living in squalor in a cold garage at the operator's home. Dogs found there were suffering from pneumonia, urine burns, parasites, and malnutrition, among other things. The operator was not registered with the State of California or the USDA and had none of the required kennel, sales, or business licenses.

Initially the operator asserted that he wasn't in business and that the dogs were his pets; this was debunked by the existence of retail customers, advertisements in local papers, and evidence gleaned from his bank accounts, pursuant to a search warrant, that showed substantial income made from the sale of puppies. He then claimed that he was not a breeder and that he only sold dogs for other breeders, which, if true, would have removed him from the jurisdiction of California's Pet Breeder Warranty Act. This didn't work, since officers could plainly see evidence of breeding. He then said that he was a breeder and therefore not subject to the husbandry, reporting, and sales regulations under the Pet Protection Act because those are for retailers and the dogs found at his home were his pets. This was also an unsuccessful argument because it was proven that he had retail customers. In fact, spcaLA had known from the start that he was a seller because its informants were customers. Of course, the animal cruelty laws and these other laws applied, so he was grasping at straws.

Twenty-six puppies were seized, all suffering from maladies that put them at death's door. Veterinarians confirmed a host of medical conditions that included dehydration, worms, uncomfortable scabs around their rears, diarrhea, wiggling teeth due to under calcification, and pneumonia. The operator had not provided any veterinary care for the puppies, had not delivered required documents when selling the puppies, had not maintained required health records, and had lied about the puppies' origin, all in violation of California's new Pet Protection Act and the Pet Breeder Warranty Act, which had just gone into effect on January 1, 1996. The operator had also refused to refund money for returned dogs or reimburse customers for medical bills that were reimbursable pursuant to the law. spcaLA officers substantiated violations of all these regulations under the new laws and further demonstrated that the evidence amounted to felony violations of the California Penal Code suite of laws, which prohibits cruelty to animals.

The operator was charged with 126 counts of animal cruelty violations and one charge of theft. (The theft charge was because he took funds from people under false pretenses and misrepresented material facts in advertisements and to buyers. Since he lied, the purchasers paid his price, and this is considered larceny under the law.) He ultimately pleaded guilty to three of the felony counts, was sentenced to three years formal probation, was permanently enjoined from being involved in the pet business and owning or being around animals, was ordered to pay restitution to spcaLA, and consented to unannounced reasonable searches and seizures of his properties.

In April 1996, for the first time in California, the district attorney also instituted concurrent causes of action under the new Pet Protection and Pet Breeder Warranty Acts that compelled the operator to make restitution to the customers who came forward once the case was publicized. He paid in excess of $185,000 by the time the case was closed. The case of Margaret and Al Valdez and their Rambos would have been resolved very differently if this law had been effect at that time.

The Hawaiian Gardens Case

In 2007, spcaLA was involved in a large puppy mill case in Hawaiian Gardens, a city in Los Angeles County. The investigation that led to the case also began after spcaLA received multiple calls from people who bought puppies who either became seriously ill or died within days or weeks of purchase. The puppy mill billed itself as a kennel and a pet shop, but it was actually a breeding facility. It imported breeder dogs and sold offspring directly to consumers and retail outlets. The operation had no valid permits for boarding, breeding, or selling. In fact, it was listed by the office of the California secretary of state as a suspended corporation, and as such it had no authority to do business at all.

A visit by spcaLA officers revealed puppies who were about two weeks old and a mixture of designer breeds, including miniature pinscher, Lhasa apso, dachshund, Maltese, poodle, Korean Jindo, and English bulldog. The person in the store admitted there was no veterinary care, but said the owner of the store vaccinated and dewormed the puppies when they were sold. It was unclear from the conversation whether underage animals were for sale or not, as the person was evasive in answering the officers.

Dogs are considered underage, and therefore not legally permitted for sale, if they are younger than eight weeks old. The eight-week requirement allows puppies to complete their weaning process. Yet ripping unweaned puppies away from their mothers, which is cruel and unhealthy for both mother and baby, is a very routine practice in the puppy mill industry, as younger dogs are considered cuter and thus more desirable.

The preliminary visit was followed by one made by spcaLA undercover operatives who purchased two puppies, a pug and a dachshund, all clearly underage. In the presence of the undercover officer, the owner vaccinated the dogs, performed skin dips (medicated baths), and dewormed them while a minor, a child, filled out the veterinary record and backdated the inoculation records. The child peeled the labels off the vaccine vials and stuck them onto the medical reports.

As the Los Angeles Department of Animal Care and Control had also received complaints about the sale of sick puppies at that location, they launched a joint operation with us. A search warrant for the entire premise was promptly obtained, and on executing the warrant and searching the scene, dogs were discovered everywhere. Though most were confined in something, two were loose—one with obvious neurological issues and the other with a prolapsed uterus. (Both were immediately sent to an emergency animal hospital.) While we were there, a customer arrived with four dead pug puppies in a box. He was seeking a refund as he had just bought them. We seized those, sent them for necropsies (animal autopsies), and took his statement.

We found more than 250 rabies certificates that appeared to have been stamped by a veterinarian, but when we spoke with the vet, he insisted that he did not provide that service. He stated that he had sent letters requesting that the owner cease and desist using his name and forging rabies certificates. This matter also required an investigation, as it was not immediately clear who the culprit was—the vet, the puppy mill operator, or both. (There are some veterinarians who earn extra income by being complicit in puppy mill operations and in puppy trafficking and smuggling operations.)

All the dogs at the location were in poor health. There were some whose coats were stained yellow from lying in their own waste for prolonged periods of time. In the breeding area, there were dogs with ulcerated eyes, lethargic and unresponsive dogs, a dead four-week-old puppy wearing a tag marked "Sold—pickup Sunday," and nursing mothers in cages suckling puppies, both dead and alive. Wire cages were stacked in the breeding areas and crammed with animals. The odors, from the birthing secretions, dead puppies, filth, ammonia, feces, mucous, and puss, were awful.

The puppies who were alive were silent.

We seized all the dogs, dead and alive, 367 in total, and every one was in poor condition. The following is an excerpt of the medical report prepared after necropsies of the dead animals were performed. The SS number refers to an in-house identification system.

SS# 1565
This was a female Dachshund that had large amounts of hard green tarter on all teeth and inflamed gums. She was neurologic, ataxic and unable to stand on physical examination. Necropsy results showed she was also suffering from alveolitis, neuronal degeneration and necrosis, pneumonia and gingivitis.

SS# 1584
This was an intact male cocker with severe skeletal muscle wasting and flea infestation. All teeth were coated in hard green tarter and the gums were inflamed. This dog presented wearing a metal collar with brass tag # 170 that was clearly too tight for its neck with resulting secondary hair loss and tissue damage. He was emaciated as well. Necropsy results showed he was suffering from pulmonary mycosis, enteritis, colitis, neuronal degeneration and necrosis, skeletal muscle atrophy, external parasitism and severe gingivitis.

SS# 1620
This was an approximately 3-year-old male yorkie that had severe and generalized crusting, alopecia and erythema. He also had retained maxillary canine teeth. All teeth were covered in large amounts of hard green tartar and his gums were red and inflamed. Necropsy results revealed he was suffering from pyogranulomatous dermatitis, dermatophytosis and gingivitis.

SS# 1625
This was a female intact cocker spaniel that was about 5 weeks old. She had discharge from her eyes and nose and was dehydrated. She also had diarrhea. Necropsy results showed she was suffering from bronchointerstitial pneumonia, Distemper virus, adenovirus, enteritis, colitis, coccidia, bone marrow atrophy and diarrhea.

SS# 1626
This was a 5week old female cocker spaniel that had discharge in both her eyes and nose. Necropsy results showed she was suffering from Bronchointerstitial pneumonia, Distemper virus, adeno virus, enteritis, colitis, coccidian, and lymphoid atrophy.

SS# 1645
This was an intact female approximately 7-year-old pug. She had severe thick green tarter on her teeth, abscessed teeth, and a large mass on her abdomen. Her lung sounds were increased as well. Necropsy results revealed she was suffering from uterine cysts, dermatitis, endometrial gland hyperplasia, gingivitis and abdominal herniation.

SS# 1820
This was an intact two-year-old female Shi Tzu. She had discharge from her eyes, nose and ears. Necropsy reports revealed she was suffering from Distemper virus, Arteritis and thrombosis, epicarditis, enteritis, conjunctivitis, giardia, otitis externa and pulmonary edema.

At the time of the raid there were several unattended children on the property, and we requested that child protective services respond to assist them. The presence of children in an environment where animals are abused, such as dog trading days, dogfights, and puppy mills, is disturbing. Desensitization to the suffering around them does not augur well for children becoming kind and empathetic adults. Judith Reitman's article on trading day in Poplar Bluff, Missouri, begged the question of how the boys who witnessed trading day would be affected by it. In the Rambo case, the family was worried about sending the right signals to their children about responsibility, care, and commitment to a pet. In this puppy mill bust, children were left in squalor with the puppies and asked to participate in the abuse. When adults, whether parents, teachers, or other mentors, condone animal cruelty, neglect, and callous indifference to suffering, we all pay the price when that violence is passed forward.

The district attorney's office filed a complaint against the operators, asking for a permanent injunction, civil penalties, and other causes of action based on violations of the California Business and Professions Code, the Health and Safety Code, and other legislation, the foundation of which was conduct violating animal cruelty statutes. A pretrial settlement was reached in which the defendants were enjoined from operating any animal-related business, required to pay in excess of $83,000, which included restitution to spcaLA, and also to pay restitution to purchasers—including refunds and medical costs incurred from the purchase of puppies. A claims administrator and public telephone number was set up at the cost of the defendant so that customers, known and as yet unknown, could file claims. Most of the dogs were rehabilitated and placed for adoption. In these situations, there is no way to be certain what medical issues will develop. All known health information is given to the adopter.

A similar case, with similar causes of action, was filed in 2015 against a chain of five pet stores. A twist in this case was that the judge granted a preliminary ruling to consider allowing the case to go forward as a class action suit, which essentially would have allowed any California customer to be a plaintiff. However, in 2017 the judge denied the motion for class action status as there were too many individual factors and variables to properly fit into a class, a decision that was a win for the stores. However, as of this writing, four of the stores have closed for reasons related to pressure from demonstrators, negative publicity, and legal issues, and the last will be closed at the end of 2018.

The Hawaiian Gardens Spinoff Case

It wouldn't be Hollywood without a spinoff!

In 2007, not long after the Hawaiian Gardens case, spcaLA officers received a tip alleging mistreatment of animals in a pet shop in Lynwood, a city in Los Angeles County. Included in the allegations were statements that live animals were left to die in a dumpster behind the store.

Our investigation revealed a horrific scene. The puppies found were mostly purebreds from at least six designer breeds, including English bulldog, Yorkshire terrier, cocker spaniel, miniature pinscher, and golden retriever. There were other assorted animals as well, all lacking food and water and covered in feces. The puppies were all suffering from anemia, malnutrition, bloat, parasites, dehydration, nasal discharge, crusty eyes, long nails, and poor teeth. Left untreated, or inadequately treated, they would die.

The pet shop had what appeared to be a medical area that was filthy, with bloody newspaper over much of the table and in the trash, as well as syringes, peroxide, and bottles of pentobarbital, a sedative that is a controlled substance. Two more dogs, pit bull puppies, were found in great discomfort, with bloody ears that had been recently cropped (cut off). A golden retriever puppy had an infected docked tail (also cut off) in addition to other problems. No anesthesia had been used in the cropping and docking. No medical care had been offered to remedy the resulting painful infections. The severed ears were found in the trash.

Assorted expired business permits and about two hundred rabies vaccine certificates were found in the shop. The certificates appeared to have been signed by a veterinarian, the same one from the Hawaiian Gardens puppy mill, but the vet again denied providing the service and said that on two occasions, he had demanded that the owner of the shop stop using his name. There were no papers showing the origin and source of the puppies, which were legally required to be on the premises, and it became clear that the owner of the shop was practicing veterinary medicine without a license, forging rabies certificates, and in possession of controlled substances that are illegal to possess without the required license from the Drug Enforcement Administration (DEA), a federal law enforcement agency.

It was soon determined that the shop had been receiving puppies from the Hawaiian Gardens puppy mill. Given the condition of the puppies there, it was clear how the misery factor kept compounding for these designer puppies. They were malnourished, sick, infested, and debilitated before they arrived at the pet store and their conditions worsened there. The store owner saw the condition of the puppies upon arrival, but did not help them. As he saw it, this was not aberrant behavior, but rather adherence to the business tenet of investing nothing in animals in order to maximize profit.

The investigation later revealed that when the store owner no longer wanted the animals, usually because they had medical issues, or they aged and were no longer puppies, he suffocated them, beat them, or threw them against the wall, and then either left them to die in a dumpster or, in at least one case, turned the dog loose on the street.

The store owner was indicted on multiple felony counts and eventually sent to state prison to serve a six-year sentence. Fortunately, we were able to successfully treat the puppies and they were ultimately adopted.

This case is again illustrative of the refrain that self-perceived success in the designer dog industry usually means minimizing investment and overhead to push profit. The behavior of the store owner was not unlike the conduct of people operating puppy mills, not surprising since they have the same goal.

It is important to note that animals like the ones found in this case and ones in similar circumstances frequently have ongoing medical problems, genetic issues notwithstanding, for the rest of their lives. Sadly, the puppies were likely born ill, the progeny of overbred and malnourished mothers, and survived stressful transport to the pet shop only to be treated as described. There are heartbreaking times when we rescue puppies, but what they've been through has been so brutal that medical treatment is often not enough and they don't make it. Other times, however, we are able to rescue puppies and, as in this case, find adopters who know of the possible problems if they take one of the puppies, but are willing to do it anyway.

If hearing all the heartbreak that goes along with the designer dog industry doesn't make you pause, I don't know what would.

The Los Compadres Swap Meet Case

Toward the end of 2011, spcaLA officers responded to a tip that animals located in a pet shop at the Los Compadres Swap Meet were sick, vomiting, and had bloody stool, and that there were dead dogs in the trash behind the shop.

Upon arriving on the scene, the officers saw a Chihuahua puppy, later named Milagras by spcaLA staff, lying on the ground near the entrance to the shop. She was lethargic, unresponsive, and breathing shallowly. There were feces coming from Milagras's anus, and she was paddling her legs in distress. She appeared to be dying. Milagras was rushed to a critical care veterinary hospital while the investigation continued.

The shop was closed to the public that day and the electricity was off, leaving the animals in extreme heat and darkness with no one looking after them. A blue storage container held three Chihuahua puppies who were covered in their own feces and urine. The food and water bowls were full of the same. Further inspection revealed wire crates with two terrier puppies and a cocker spaniel puppy. The crates and food bowls there were covered in feces, as were the puppies. One of the crates had a water bottle with the spout facing outward so the dogs could see the water but not access it. There did not appear to be breeding apparatus at the location, and it was confirmed that the dogs were purchased elsewhere. The operators had no origin paperwork for the animals and no veterinary documents chronicling prior care. When asked, the owner of the establishment said that no name of an emergency vet "came to mind." What follows is an excerpt from Milagras's veterinary report that notes her condition:

critical on admission due to dehydration, malnutrition, diarrhea, parasitism and hypoglycemia and would have died if she was not immediately treated. The likely cause for her physiologic state was neglect from inadequate husbandry. Without adequate food, water and too high of an environmenstal temperature this puppy was severely neglected.

The puppies in this pet shop were purebreds ordered from breeders. If they were ill upon arrival at the pet shop or became sick while there, they were not provided veterinary care and were instead trashed. No funds were spent on sick puppies. Necessities like electricity, food, water, and care were doled out in miserly amounts and not provided at all when the shop was closed to the public. When puppies died or were disposed of, the price of the survivors was adjusted to cover the cost of the initial order: minimal investment to maximize profit.

In this case, all the puppies were successfully treated and ultimately adopted after spcaLA's investigation. The shop owner was convicted of animal cruelty, placed on probation for three years, ordered to undergo counseling, ordered to pay restitution of approximately $8,600, and prohibited from owning or having anything to do with animals for five years.

I've chosen to describe this case and the others since they are excellent, though tragic, illustrations of the pain dogs suffer, from the puppy mill to the pet shop—a picture-perfect microcosm of the industry. Born in squalor, transported in pain, then left in wretchedness waiting to be bought, with no relief, no care, and no peace, at only, on average, several weeks old. Is this the cost of cute?

If you think that it is basically the same sad story as the 1989 Rambo story told over and over again, you would be correct.

The only difference is that there are more laws today than there were in 1989 to punish the practice once it is discovered and successfully prosecuted. However, the atrocity of what happens in most puppy mills and pet shops is rarely discovered, so the incentive to reform is low. The odds are against getting caught.

And what about the buyers? Do those who purchase these high-priced dogs get their money's worth?

CHAPTER 6

WELCOME TO THE US OF A

The question is not, Can they reason? Nor, Can they talk? But, Can they suffer?

—*JEREMY BENTHAM*

It is difficult to locate the designer dog puppy traffickers in the United States, and even harder to locate those involved in worldwide trafficking. Puppy mills and unethical breeders all over the world will ship puppies to anywhere there is a desiring buyer with the cash to pay. Buyers can be other puppy mills, laboratories, pet stores, or any point-and-click purchaser. In June 2018, Fox News, citing the National Animal Interest Alliance, reported that Americans are importing approximately one million animals a year from Turkey, the Middle East, China, and Korea.[19] Fox characterized the United States as a "dumping ground" for foreign dog imports. These are the ones that are known, rather than smuggled. It is a disgrace when healthy dogs in America are euthanized because they're considered surplus.

---

19 Sheila Goffe, "The US Has Become a Dumping Ground for Foreign 'Puppy Mill' and 'Rescue' Dogs. Here's What Needs to Change," Fox News (June 30, 2018), http://www.foxnews.com/opinion/2018/06/30/us-has-become-dumping-ground-for-foreign-puppy-mill-and-rescue-dogs-here-s-what-needs-to-change.html.

Pursuant to the US Customs and Border Protection Regulations for US Residents, underage, unvaccinated, and sick animals may not enter the United States. The minimum age requirement has changed from three months to six months, but the idea and message are the same:

> Dogs must also be free of diseases that could be communicable to humans. Puppies must be confined at a place of the owner's choosing, which can be a private residence, until they are three months old and then they must be vaccinated against rabies. The puppy will then have to stay in confinement for another 30 days following the vaccinations.

> Dogs older than three months must get a rabies vaccination at least 30 days before they come to the United States and must be accompanied by a valid rabies vaccination certificate if coming from a country that is not rabies-free. This certificate should identify the dog, show the date of vaccination, the date it expires (there are one year and three year vaccinations), and be signed by a licensed veterinarian.[20]

Based on a multitude of consumer complaints, it is suspected that thousands of underage designer puppies, including toy breeds and teacups, are transported into the United States each year. The complaints allege that many of these dogs are so weak from their stays at puppy mills and the awful conditions they travel in that they die within forty-eight hours of purchase despite high prices (which vary with the breed, but can be in the thousands of dollars) and gargantuan veterinary bills. There are also a wealth

---

20 U.S. Customs and Border Protection, Know Before You Go: U.S. Customs and Border Regulations for U.S. Residents (July 2006).

of complaints from people who purchase underage and very sick dogs from people at swap meets and informal meeting spots, like parking lots and street corners, as well as an uptick in puppies brought to shelters while suffering from parvovirus, distemper, and other ailments. Many of the purchasers were told the puppies came from puppy mills in Mexico.

When dogs are transported from one location to another, it's hard to know if they will bring diseases with them. Diseased dogs could infect other canine populations and even humans and cause canine or zoonotic outbreaks. An example of a canine outbreak followed Hurricane Katrina, when dogs were helped by animal rescue organizations throughout the country. spcaLA, for example, was asked to accept an airlift of hundreds of stranded and injured storm refugees. Knowledgeable organizations like spcaLA instituted protocols to avoid the spread of diseases that the Hurricane Katrina dogs could bring, and to give arriving sick dogs the best care possible. Heartworm was one of our major concerns because the Hurricane Katrina dogs were coming from a warm, humid climate where heartworm is prevalent.

Other individuals and organizations, some well-intentioned and some eager to monetize and publicize "hurricane dogs," didn't utilize disease prevention protocols, and inadvertently allowed heartworm to spread from one dog to the next. Some of the Louisiana dogs were adopted, purchased, or fostered by people who already had dogs in their homes, and when their dogs were infected an epidemic of heartworm was born. Other Hurricane Katrina dogs were placed in shelters where they infected other shelter dogs, or were left on the street and infected other street dogs. Exposed dogs were then also adopted, purchased, and fostered, and the spread of the disease worsened. Today, dogs throughout California have heartworm.

With proper precautions, this could have been avoided.

Other ailments that many of the Hurricane Katrina dogs arrived with were ringworm, scabies, and giardia. I recall the time my staff and I stood outside the plane carrying the Hurricane Katrina rescues we were receiving; we were all gloved and gowned. To our horror, when the plane door opened, we saw one of the airline staffers exit the plane carrying a dog whom she then kissed on the mouth. The staffer wasn't thinking about how the dog had been pulled out of bilge water loaded with contaminants from destruction debris and was traveling in close quarters with hundreds of other dogs similarly exposed. She also wasn't thinking that she could reasonably expect to become ill with whatever disease the dog she kissed carried or others whom she may have played with on the plane carried. The staffer's show of affection was well intentioned, but reckless.

We are now experiencing a similar situation as people unknowingly bring dogs with canine influenza from China into the United States. The dogs have introduced a strain of canine influenza that hadn't previously existed in California. Different disease, same story.

Border regulations are supposed to prevent the transmission of diseases between countries. Some states have passed their own regulations in an attempt to prevent diseases from entering and then spreading within their state. In 2014, California passed a law requiring that within ten days of a dog being imported into the state for resale a health certificate be presented to a county health department.

The Border Puppy Task Force

In 2005, spcaLA, along with a consortium of fourteen other organizations and the cooperation of the federal government, formed a task force that positioned itself at the California/Mexico border. The mission of the task force, which we named the Border Puppy Task Force, was to study the entrance of infected animals into the United States and assess the gravity of the entrance of diseased animals, the conditions the animals travel in, and fraudulent business practices surrounding their entrance. Our plan was to have an initial seven-day, twenty-four-hour surveillance operation, and our list of questions included: What breeds of dogs are coming in? Are the dogs underage? Have they been vaccinated? Do they have canine and zoonotic diseases? Is the method of their transportation humane? And why are they coming? We hoped that answering these questions would inform our ability to create solutions.

We set up shop at both the San Ysidro and Otay Mesa border crossings between Mexico and San Diego. On a normal day, people can walk or drive into the United States. The pedestrian queues are long and the vehicle lines can take hours, and anyone, American citizen or not, and any vehicle, can be searched. The searches can be at random, due to law enforcement intelligence, due to an indication by a trained drug or explosive detection dog, or due to the "Spidey-sense" of an officer, his or her intuition. A search of a vehicle can be as cursory as "Hello and welcome to the United States, please let me take a quick peak inside your car," or as involved as the car being completely dismantled down to its tires and chassis, if an officer feels something is amiss. Because of the length of time it takes to get into California, it's especially important that people are prepared to cope with the climate at the border, which ranges from warm to extremely hot, and prepare any animals they carry with them to do the same. The heat can be dangerous, even deadly.

The Border Puppy Task Force's focus was on the vehicles rather than the pedestrians entering the United States, as it is more likely for dealers and breeders to smuggle in larger numbers of dogs by vehicle than by carrying them on the body. That said, one puppy in a backpack could fetch a few thousand dollars, which can make a walk good exercise and profitable.

Working with US Customs officers, the Centers for Disease Control and Prevention (CDC), and the California, Los Angeles, and San Diego Departments of Health, the task force was given permission to investigate and inspect any vehicle with animals in it who were clearly visible from the outside, when there was a suspicion it held animals, or when animals were found pursuant to a search.

What we discovered was eye-opening. The large number of dogs that people attempted to smuggle across the border was remarkable. Most were custom-designed dogs, primarily cockapoos, maltipoos (a cross between a Maltese and a toy or miniature poodle), Lhasa mixes, Chihuahuas, and terriers. Many were underage, and many were tiny. Some were so young that they didn't have their eyes open yet. Papers were falsified, rabies vaccine certificates were provided for dogs clearly too young to be vaccinated, and destination points included fictional addresses. If they weren't already in terrible condition from the puppy mills when they began their journey, they were by the time we saw them, as a result of the transport conditions.

I have soul-searing memories of things I saw at the border. Puppies, several weeks old, were stuffed into hollowed-out rear seats, auxiliary gas tanks, small suitcases, toolboxes, glove compartments, and tire wells. They were crammed in, the dead with the living, and they were suffering from all the illnesses common to puppies bred in deplorable conditions, compounded by heat stroke, dehydration, and travel-related injuries. They were caked in their own waste and their limbs were twisted unnaturally by people cramming as many into a container as possible. One dog's mouth was stuck shut by her own vomit.

The puppies were also forced to remain confined for hours as they traveled to the border and then waited in line to enter the United States. Because of the long wait time, the drivers would frequently turn off their cars to save gas. They would step outside, drink water, and chat while the puppies were left in the cars without air conditioning or water.

If you think that puppies are only exploited for sale profits, think again. They are also purchased by drug dealers and used to smuggle in drugs under the guise of being part of the pet trade. In one memorable instance on the East Coast, the DEA arrested twenty-two Colombian nationals for smuggling heroin into the United States by surgically implanting liquid heroin into the bellies of purebred puppies. The investigation involved $20 million worth of heroin and employed people who swallowed or surgically implanted the drug into their bodies as well as into the puppies, planning to remove the packets once the carrier cleared customs.

Fatalities can occur from performing surgery on puppies so young, as well as from heroin leaks that cause overdoses. These are just viewed as mishaps by the smugglers, part of the cost of doing business. The bags that don't leak can be recovered. The puppy need not remain alive once in the United States for the perpetrators to deem the operation a success. The enormous amount of worldwide trafficking of these puppies, coupled with poor enforcement, provides an easy opportunity to move contraband.

With a plan to have a designer dog ingest a designer drug, what could go wrong?

When animals were found at the border, documents were checked and the necessary action taken. Depending on the findings, the people and animals were admitted into the United States, detained, or sent back to Mexico; some people were sent back alone, their animals seized by the Border Puppy Task Force. While interviewing detainees we were told, "People want these dogs, they will pay a lot for them, and we will provide them." We found out that there were puppy mills not far from the border that were churning out cockapoos and other small-breed dogs in appalling and cruel conditions.

If people and their animals were admitted into the United States, a judgment made by officers at the border, then data collected at the border was entered into a data bank and used to support follow-up investigations to determine if quarantine and confinement orders were obeyed, if humane conditions were provided, and if sales were conducted in keeping with the law. Questions for follow-up investigations included: Were diseases coming in? Was fraud occurring? Were laws broken after entering the country (i.e., were underage and/or sick puppies being sold illegally once they arrived in the United States)? This intelligence led to sting operations that caught people selling underage and/ or sick dogs in violation of the law.

Since its initial work at the border in 2005, the Border Puppy Task Force has returned to the border on random occasions at random times for quick checks, including just before Christmas, which is a popular time to gift a puppy. As is typical, the puppies bought for Christmas were often dead or abandoned before New Year's Eve.

The Border Puppy Task Force saw more than thirty-five

hundred dogs. Almost one-third were under three months old, and about one-half were under six months old. (We also checked horses, cats, hamsters, and parakeets.) Of that number, almost one thousand were quarantined for lack of rabies vaccinations and almost one hundred were seized and rushed to emergency veterinary hospitals in the area. Of those, two were euthanized. The youngest dog seized was a two-week-old Chihuahua. The rest were either turned back to Mexico or allowed to cross into the United States as their papers were in order and the dogs were healthy. Most of the latter were the dogs of Americans and Mexicans who were traveling with their family pet. Typically, their vehicles had only one or two dogs. When we found multiple dogs in one vehicle it was because a breeder or seller was transporting them.

Imagine the number of animals who would be found if there was a dedicated effort to conduct these investigations throughout the year. This could be done for brief stints as often as possible. More criminals are caught when operations are conducted at random intervals, rather than during a sustained stay. Once word got out that we were inspecting vehicles, the transportation of animals thinned, but surprise follow-up visits showed that attempts to bring animals into the country always picked up again. We can't set up permanently, not for a lack of wanting to, but rather a lack of resources. There is a great need for this specialized enforcement.

The information gathered by the Border Puppy Task Force provided a lot of "puppy pipeline" intelligence for follow-up law enforcement efforts; tracing the paths of the dogs and repeat border crossers yielded results. For example, spcaLA and one of our task force partners conducted a sting operation in a neighborhood parking lot. An undercover officer posing as a consumer interested in purchasing a puppy arrived at the parking lot carrying (marked) cash only, as instructed by the seller, to purchase underage puppies who had been smuggled in from Mexico. After the buy took place, the seller was immediately arrested. He was charged with selling puppies under eight weeks old, a misdemeanor in California. Intelligence gathered from the defendant cross-checked against the Border Puppy Task Force's information led to more sellers and more arrests. Each one of the sellers gave us information about his or her colleagues. Sellers should worry that the person on the other end of the phone or email, answering an advertisement, is one of us. We can operate in secret too.

The information gathered was also used to support the enactment of laws in California that prohibit suppliers from entering the state, and protect the puppies they attempt to import. As a result of tracking where these suppliers were selling their puppies, and then filing charges, California chose to make it illegal to sell dogs on the streets and in public places. Thanks to data gathered by the Border Puppy Task Force, spcaLA recently convicted one individual on two counts of violating the California Penal Code—selling a puppy under eight weeks and selling said puppy in a public place. It is encouraging that the operations of the Border Puppy Task Force yielded new prophylactic state laws designed to hinder these sales; they also contributed to the federal effort to clamp down on internet transactions. A little bit of law enforcement can go a long way, but only if resources are allocated for it.

In 2006, the Border Puppy Task Force received international attention when Congressman Bob Filner presented it with a "Crime Fighter of the Month" award. He stated:

The Border Puppy Task Force's leadership and efforts are truly inspirational. Their commitment to continue the fight against such inhumane treatment of man's best friend is highly commendable and I am honored to present the team with this award.

The inhumane transportation of dogs across the border persists. In April 2016, twenty-four puppies were found unresponsive in a hot car at the San Ysidro border crossing. The puppies, a mix of Labradors, shepherds, terriers, and Chihuahuas, appeared dead, but once they were given fresh air, some revived and the others were raced to an emergency veterinary hospital. The driver said the puppies came from Tijuana (right across the border) and that before they left they were given water. By the time they were discovered, the slow traffic had caused it to be five-and-a-half hours since that drink—too long an interval for anyone, let alone a puppy. The driver was charged with twenty-three counts of animal cruelty. The puppies were cared for by our friends at San Diego Animal Services and made available for adoption.

The latest trend is for unethical breeders and middlemen to pretend to be members of nonprofit animal rescue organizations. Some actually incorporate and file the appropriate documents; others just lie about it and solicit donations anyway. A good example of this is the recent arrest of an owner of a fake nonprofit who established an online presence and knowingly sold sick and underage puppies to the public. It is alleged that these puppies were smuggled into California from Mexico. The initial complainant bought a very sick puppy who died within a week of purchase, after intense and expensive medical intervention failed to help. More customers with similar complaints were discovered on Yelp and others shared their stories in response to news coverage. The seller was arrested with six underage and very sick puppies in her possession and we found thirteen more at another location. At the time of this writing, law enforcement is investigating whether the owner is operating under other fabricated businesses and nonprofit organizations. She was arrested and charged with multiple felony and misdemeanor counts, including animal cruelty and assorted violations of retail and business law.

Hiding behind a nonprofit can allow puppy mill operators to circumvent laws designed to keep commercial sellers (pet shops) out, but allow charities in. When they can sidestep retail source bans, the fake nonprofits manipulate families into believing they will be helping a nonprofit dedicated to rescuing homeless and abandoned animals if they purchase one of its dogs. In June 2018, the *Chicago Tribune* detailed this practice of commercial dealers forming their own not-for-profit "rescues" in order to supply local Chicago stores with out-of-state puppy mill-bred dogs.[21] (A Chicago ordinance requires that dogs sold in pet shops must come from rescue organizations.) The *Tribune* calls it "puppy laundering." I concur with that description, as it is both accurate and connotes and comports with the criminal element in pet trafficking.

I have been criticized for not being sympathetic to people who buy a designer dog from places like the back of a pickup truck in a parking lot. These puppies often don't survive forty-eight hours and the buyer then realizes that he paid a couple thousand dollars in cash, which seemed like a bargain, to a guy with a truck full of puppies, a burner phone, and a smile. The buyer and the

---

21 Editorial Board, "Puppy Laundering in Chicago," *Chicago Tribune* (June 8, 2018), http://www.chicagotribune.com/news/opinion/editorials/ct-edit-puppy-mills-chicago-rescue-20180529-story.html.

buyer's family are devastated, and my first response is to berate the person for his lack of common sense.

If someone on the streets of New York City offers to sell you a Rolex watch for five dollars, two things are true: it will not tell the correct time, and you will have no one to blame but yourself for the loss of five dollars. Similarly with dogs, foolishness has consequences. A $3,000 designer puppy offered under illegitimate circumstances and at a "discount price" will often be ill, possibly even on death's door, will have forged papers, and could cost you $10,000 in immediate lifesaving medical bills. Of course, no returns are possible. The seller is usually long gone after the transaction, leaving you with a sick or dead puppy, no recourse, and a lot of heartache. It's nothing but a con game.

There are more of these schemes and crimes than we have the resources to deal with. I continue to tell consumers that the way to stop this is to stop buying dogs from these people. After all, how much fun is it to have your Christmas present die in your arms and to be complicit in the suffering of these animals? The callous trafficking of these puppies hurts all of us and is a problem we can correct, but choose not to. People don't like me for saying this, but my sympathies are with the dogs.

CHAPTER 7

HOW MUCH IS THAT DOGGIE IN THE 747?

Woe to the man who seeks to shed a brilliant light in a place where people want to keep in darkness and shadow.

—BENEDETTO CROCE

When designer dogs arrive at an airport that accepts animals shipped internationally, they are often very sick. This puts authorities in a tough spot. They can refuse the shipment and send the dogs back to their country of origin, which can cause another twenty-four hours of crate travel, often in the dogs' own waste and/or with dead companions, or they can seize and euthanize them, because they can't allow diseased animals into the United States. After all, they could transmit canine and/or zoonotic illnesses. The required paperwork accompanying the animals is gibberish, forged, lacking important components, not in English, or clearly fiction, since the descriptions on the manifests don't match the dogs in the crates.

It is not uncommon for a seller's agents to get spooked and abandon the dogs altogether if they see law enforcement activity near them. It is common for shippers to arrange flights that land at off-hours, on weekends, and in small airports where they think inspection agents won't be on duty. There are also people who try to disguise what's inside the crates to circumvent inspections altogether. It is not unusual for animals to be stuffed into a crate marked "coffee." Illegal, but not unusual. For fear of liability and bad publicity, many airlines will not knowingly carry live cargo.

Another legal wrinkle is that international airports, though located in the United States, suffer the problem of cargo not technically being in the United States until it has cleared customs and immigration controls. Therefore, state and federal laws don't apply until the shipment has been allowed in. This presents questions of who has jurisdiction and what laws apply after the shipment has departed the plane but before it has cleared customs, and, given that the laws don't provide immediate solutions in the best interests of the animals, the dilemma of what to do with them that anyone with empathy would feel at the sight of the contents of these shipments.

In early 2013, the Importation Task Force, which was the next version of, and has subsumed the Border Puppy Task Force, was created to work on counteracting animal trafficking and smuggling, typically via air. spcaLA is a member of this task force; other members include the Los Angeles County Department of Public Health, veterinarians, enforcement personnel from the USDA, CDC, US Customs, and other animal welfare and law enforcement organizations, such as San Diego Animal Services, one of our key partners in the Border Puppy Task Force. Members of the Importation Task Force work together and share intelligence to stop illegal shipments of puppies and follow up on illegal activity involving the puppies once a shipment is cleared. Simultaneous with the formation of this task force, changes to federal regulations regarding the importation of live dogs and the new federal definition of retail pet stores to include point-and-click sales were in draft stages. Both were scheduled to become law in late 2014.

Welcome to Ukraine, Los Angeles

In early 2014, a shipment of French bulldogs arrived at Los Angeles International Airport (LAX) from a puppy mill in Ukraine. The dogs were examined and found to be underage, not vaccinated against rabies, and in urgent need of care. The paperwork did not match the puppies in the container.

The CDC officer on duty was confronted with the ethical dilemma of whether to send the shipment back, knowing the dogs would not survive the long flight back to Ukraine, or seize them and euthanize them as they were not safe to admit into the United States. He couldn't bring himself to do either and instead contacted the Importation Task Force and turned the puppies over to spcaLA for care. However, he made it clear that he was technically refusing the dogs' entry into the United States, and that the part of our shelter that would house these dogs was to be considered in Ukraine territory. We were required to hold the puppies until they were well and vaccinated against rabies, and for thirty days after that—so several months. This seemed like a reasonable solution, but it caused a major kerfuffle within the government and for the carrier. You would think declaring our shelter as residing in Ukraine would be the issue, but in fact, funding was the problem. Caring for those puppies was expensive and covering this cost is legally first the responsibility of the airline that accepted them, and then ultimately that of the seller or shipper. In other words, the airline was on the hook and did not appreciate it.

The good news is that the episode allowed for a nationwide conversation involving our Importation Task Force, the airline, and all the respective lawyers, and we were able to discuss certain common-sense solutions for tightening enforcement in airports. Suggestions included regulating and limiting which airports can accept animals in the first place, and limiting such arrivals to certain hours to help with the problem of resources. We also discussed the need for consistent procedures should the puppies be too sick to survive the trip back, but capable of being helped if medical care is administered quickly. In the Ukraine case, the airline paid the bill, the puppies recovered, and they were ultimately released for adoption. Proposed new laws designating specific airports to accept flights with live cargo and a limited range of arrival times for such cargo have not yet been adopted, but the laws that went into effect in late 2014, increasing the age of entry and internet sales oversight, did make some important improvements.

Code of Federal Regulations (CFR), sections 2.150 thru 2.153, now governs the importation of live dogs. The summary policy statement cites animal welfare as the impetus behind the regulations, one of them stating that puppies have to be at least six months old, as opposed to three months old, to enter the United States:

> We are amending the regulations to implement an amendment to the Animal Welfare Act (AWA). The Food, Conservation, and Energy Act of 2008 added a new section to the AWA to restrict the importation of certain live dogs. Consistent with this amendment, this rule prohibits the importation of dogs, with limited exceptions, from any part of the world into the continental United States or Hawaii for purposes of resale,

research, or veterinary treatment, unless the dogs are in good health, have received all necessary vaccinations, and are at least 6 months of age. This action is necessary to implement the amendment to the AWA and will help to ensure the welfare of imported dogs. *Effective date:* August 18, 2014.[22]

According to the updated regulations, the permit requirements, among other things, must include:

(1) The name and address of the person intending to export the dog(s) to the continental United States or Hawaii;
(2) The name and address of the person intending to import the dog(s) into the continental United States or Hawaii;
(3) The number of dogs to be imported and the breed, sex, age, color, markings, and other identifying information of each dog;
(4) The purpose of the importation;
(5) The port of embarkation and the mode of transportation;
(6) The port of entry in the United States;
(7) The proposed date of arrival in the continental United States or Hawaii; and
(8) The name and address of the person to whom the dog(s) will be delivered in the continental United States or Hawaii and, if the dog(s) is or are imported for research purposes, the USDA registration number of the research facility where the dog will be used for research, tests, or experiments.[23]

---

22 Department of Agriculture, Animal and Plant Health Inspection Service, "Animal Welfare; Importation of Live Dogs," Federal Register Volume 79, Number 159 (August 18, 2014), https://www.gpo.gov/fdsys/granule/FR-2014-08-18/2014-19515.
23 Ibid.

The regulations also call for certain certificates, which must be filled out in English, to document the receipt of required vaccinations and certify that the dog is free from parasites, infections, skin diseases, and nervous system disturbances, and is not emaciated, jaundiced, or suffering from diarrhea, all of which are among the top ten problems endemic to puppy mill and backyard bred dogs.

(a) *Required certificates* . . . no person shall import a live dog from any part of the world into the continental United States or Hawaii for purposes of resale, research, or veterinary treatment unless the following conditions are met:

(1) *Health certificate.* Each dog is accompanied by an original health certificate issued in English by a licensed veterinarian with a valid license to practice veterinary medicine in the country of export that:

(i)   Specifies the name and address of the person intending to import the dog into the continental United States or Hawaii;

(ii) Identifies the dog on the basis of breed, sex, age, color, markings, and other identifying information;

(iii) States that the dog is at least 6 months of age;

(iv) States that the dog was vaccinated, not more than 12 months before the date of arrival at the US port, for distemper, hepatitis, leptospirosis, parvovirus, and parainfluenza virus (DHLPP) at a frequency that provides continuous protection of the dog from those diseases and is in accordance with currently accepted practices as cited in veterinary medicine reference guides;

(v) States that the dog is in good health (i.e., free of any infectious disease or physical abnormality which would endanger the dog or other animals or endanger public health, including, but not limited to, parasitic infection, emaciation, lesions of the skin, nervous system disturbances, jaundice, or diarrhea); and

(vi) Bears the signature and the license number of the veterinarian issuing the certificate.

(2) *Rabies vaccination certificate.* Each dog is accompanied by a valid rabies vaccination certificate that was issued in English by a licensed veterinarian with a valid license to practice veterinary medicine in the country of export for the dog not less than 3 months of age at the time of vaccination that:

(i) Specifies the name and address of the person intending to import the dog into the continental United States or Hawaii;

(ii) Identifies the dog on the basis of breed, sex, age, color, markings, and other identifying information;

(iii) Specifies a date of rabies vaccination at least 30 days before the date of arrival of the dog at a US port;

(iv) Specifies a date of expiration of the vaccination which is after the date of arrival of the dog at a US port. If no date of expiration is specified, then the date of vaccination shall be no more than 12 months before the date of arrival at a US port; and

(v) Bears the signature and the license number of the veterinarian issuing the certificate.[24]

---

24 Ibid.

One of the most significant changes is that the USDA would no longer offer confinement agreements. These agreements used to allow someone without the appropriate rabies inoculations to promise to keep the dog confined at a location until thirty days after inoculation, essentially what we did in the Ukraine case. Of course, most importers don't comply with the agreement, and those locations usually don't exist. We were able to document through the Border Puppy Task Force that some of the listed confinement locations were vacant lots. It is more efficient to prevent entry in the first place than to have to find the puppies and their importer, and risk allowing rabies and other infectious diseases into the public. The CDC still allows these confinement agreements in restricted instances, but is very stingy about doing so.

Those of us on the Importation Task Force can communicate and assist one another in handling anomalies on a case-by-case basis. Will people forge documents, pay veterinarians to fill out forms, pretend their dogs fall under an exception, such as that applicable to a personal pet, and smuggle teacup animals in luggage? Of course, but we already know that and are prepared. In addition to dogs, the Importation Task Force also responds to cases of other animals, and some reptiles, whose owners attempt to bring them into the United States.

One thing that was confirmed for us was that there are people in the United States who claim to practice careful, humane, and ethical breeding, but are actually importing puppies from another country. People who buy their dogs may never suspect that their chosen dog came from a grungy puppy mill on the other side of the world. We observed precisely this scenario in a French bulldog case. At the time, we were monitoring the activities of a breeder whose facility had no obvious violations. All dogs appeared clean and properly documented until a shipment of French bulldog puppies arrived at LAX from Ukraine, destined for the breeder's location. The shipment was denied entry into the United States, since the dogs, paperwork, and inoculation records did not match up. We then learned that the puppies were under eight weeks old. After a veterinary exam, the dogs were returned to Ukraine. The breeder appealed the decision to return the dogs, but was denied. The denial letter read:

CDC has received your written appeal requesting reconsideration of CDC's previous denial of your request to import four dogs into the United States from Ukraine. After thorough review by a CDC management official who is senior to the employee who made the initial denial of your initial request, which included the reasons for the initial denial, and your written appeal, CDC has denied your appeal.

On September 10, 2014, your shipment of four French bulldog puppies arrived at Los Angeles International Airport. The shipment was accompanied by rabies vaccination certificates that stated that the dogs were 4 months of age and that they had all been vaccinated against rabies at 3 months of age. However, upon arrival at the US port of entry, your shipment was inspected by a Los Angeles County public health veterinarian on behalf of CDC. The veterinarian found that the physical appearance of your dogs is inconsistent with the stated age on the rabies vaccination certificates, and that in fact the dogs appear to be less than 8 weeks of age. As the dogs in your shipment do not match the accompanying documentation, the rabies vaccination certificates are invalid and your dogs are considered inadequately immunized against rabies under CDC's dog importation regulations.

As your dogs are inadequately immunized against rabies, CDC evaluated your shipment's eligibility to be imported under the terms of a confinement agreement. However, the number of dogs that you are importing (four), in combination with your intent to import them for resale, makes your shipment ineligible for a confinement agreement. Additionally, the prevalence of rabies in your dogs' country of origin (Ukraine) presents a high risk of importing rabies into the United States.

This denial of your appeal constitutes final agency action. Please do not attempt to re-import your dogs into the United States before they are at least 4 months of age and fully immunized against rabies.[25]

There is a particularly high rate of French bulldog shipments because of the high demand for them. CBS News reports that in just twenty years French bulldogs have risen in popularity from seventy-sixth to fourth place on the list of most popular dog breeds in the United States, a finding echoed by many surveys.[26] French bulldogs are extremely popular with the public and with celebrities—owners include Lady Gaga, Martha Stewart, Hilary Duff, Hugh Jackman, Ashley Olson, Reese Witherspoon, Ashlee Simpson, and David and Victoria Beckham, to name a few. Even if they're obtained through a reputable breeder, purchase of the dogs is expensive (each typically hovers around $3,000 to $5,000), as is their maintenance. Most people don't, however, have well-bred purebreds from reputable breeds, but instead have knockoffs from puppy mills, which ironically can end up

---

25 Retrieved from the records of the Importation Task Force.
26 CBS News, "Most Popular Dog Breeds in the U.S.," https://www.cbsnews.com/pictures/most-popular-dog-breeds-in-the-us/.

costing even more than expensive name-brand dogs from high-end breeders, because of expensive medical bills and surgeries that the knockoffs need.

These knockoffs are the offspring of parent dogs who have been bred and inbred so often that their offspring are frequently born with defects that cause them to be uncomfortable every day of their lives. The likelihood of their being born with defects is increased by the difficult birthing process. French bulldogs are usually bred smaller than they should be, so they have big heads and tiny back hips, which are a physical barrier to breeding and the delivery of puppies. They're also at high risk of overheating and going into respiratory distress during birthing since they have difficulty breathing, which is caused by their flat noses.

For the most part, the bodies of French bulldogs have been so distorted from inbreeding that they cannot even breed on their own. Mothers typically need a caesarian section, since they could not withstand the strain of labor. And after the painful birthing, the puppies are at high risk of dying prematurely. The result is that outside of the puppy-mill world there are veterinarians who specialize in artificial insemination and managing high-risk pregnancies (for French bulldogs and other breeds). However, the vets don't always have the genetic expertise or information from the owner that is necessary to avoid "breeding-in" (breeding together dogs who have the same genetic problems and will thus be likely to give them to their offspring). A healthy breeding and birthing process can be difficult and dangerous even with veterinarians in charge, and light-years worse if a puppy mill operator is overseeing it. The breeding and birthing problems that French bulldogs and other breeds have will ultimately cause them to become extinct.

One hobbyist thought it would be fun to mix two French bulldogs with different colored spots to see if puppies with multicolored spots would be born. After all, French bulldogs come in blue, black, and white, so why not try for multicolor? The result was a litter of blind and deaf white puppies who didn't live longer than a few minutes after birth. Genetic considerations are rarely considered by hobbyists and never considered by puppy mill breeders, which is why consumers don't know what they're getting. Some sellers, even if they advertise themselves as breeders, opt to order their dogs from another country to avoid the hassle of breeding; they tell themselves they are doing it for genetic diversity, but they don't share this with the buyer.

I once participated in a radio show that discussed the poor ethics of breeding French bulldogs, since they are almost always physically uncomfortable. A listener called in and said her French bulldog cost $5,000 to buy and that she pays $10,000 a year in medical bills. Other listeners called in and said they opted for surgery immediately after purchase, to help their dog breathe easier during his or her life. One veterinarian remarked that the most comfortable and peaceful that he ever saw a French bulldog was during a neuter procedure when the dog was under general anesthesia and could sleep and breathe without effort. Imagine the extreme discomfort that a dog like this would be in at a few weeks old, piled into a crate with others and traveling from Ukraine to Los Angeles.

This is the poster dog for the painfully high cost of cute.

A Teacup with Seoul

This next case has everything. It combines the trafficking of sick puppies, high prices, fake names, forged records, and internet sales.

In November 2016, a puppy broker in Northern California was sentenced for felony animal cruelty, operating without a license, and eleven other charges related to the business of selling puppies from South Korea. The sales were conducted around the world, both over the internet and in person. The puppies were underage, assorted breeds, mostly teacups, and all miserably sick. There was a high probability that the required veterinary records were forged. The puppies cost at least $4,000 each.

Numerous other puppies in the broker's possession were also in poor condition, lacking food, water, and ventilation, and visibly sick. One such puppy, just a couple weeks old, was found near death; she was extremely dehydrated, and her mouth was sealed shut from the dehydration. She did not survive. The broker was already on probation for a prior conviction of running the same kind of operation, but the conviction and fine ordered were just a nuisance to him, as they are to others like him. When these brokers are fined, they don't change their ways, they just adjust the costs of their dogs to cover the costs of the fines and still make a profit.

This particular broker used multiple business names as well as assorted aliases to evade detection and be able to shutter any business quickly if necessary. This time the broker was sentenced to one year in jail, prohibited from dealing in or possessing dogs for at least ten years, and ordered to pay restitution and other penalties. Upon discharge from jail, the broker will be on felony probation for five years. Kudos to our colleagues up north.

It's a Small World

A UK charity, Dogs Trust, carried out an investigation in 2014 to study the trafficking of puppies from Eastern Europe into the United Kingdom. In 2012, once rules were relaxed to allow people to travel with their family pets, there was a huge spike in the traffic of dogs entering the United Kingdom. The enormous rise in the number of dogs coming from Hungary and Lithuania was suspect.

As is the case in the United States, the concern was that dogs were entering the country with rabies, tapeworm, forged veterinary records, fake inoculation records, and all sorts of infectious diseases. There was also a concern that dog breeds banned because officials considered them dangerous, such as the pit bull, were being smuggled in.

The lack of resources at the border allowed smugglers to cross at odd hours and on weekends, times when there were insufficient numbers of agents on duty. Despite extensive media campaigns warning the public of the dangers and cruelties of the illegal dog trade, the demand for trendy puppies did not abate. And as in the United States, the numbers surged around the holidays. The problem was furthered by concern over the spread of infectious diseases, since the puppies were often sold at markets and congested areas where diseases can spread quickly. Additionally, officials on duty at the border were not trained to identify underage dogs and could not be expected to discern the difference between one too young to bring in and one of age; veterinarians can tell age through teeth and other attributes, but border inspectors can't necessarily do the same. There was also no sharing of intelligence among agencies who worked to prevent the entrance of illegal dogs, and no significant penalties imposed if a smuggler was caught.

UK officials work to address the smuggling of dogs by limiting entry points to the country and concentrating resources at these entries. They've also asked for a ban on the entrance of puppies under six months old and are currently creating shared databases and intelligence findings, and continuing to ask the public to stop buying dogs who are not from shelters or high-quality breeders. Officials believe that if people know that the puppies are separated from their mothers, stuffed into suitcases, and forced to travel over a thousand miles, people would not want them.

There were discussions about banning third-party sales in the United Kingdom altogether, but there is a fear that instead of ending smuggling, this would drive smugglers further underground, which would be worse for the puppies because they'd be harder to find and save. As in the United States, there aren't enough ethical breeders in the United Kingdom to meet the demand for designer dogs, so their presence raises this question: where are these dogs coming from? If someone is illegally importing designer dogs, someone elsewhere has to be exporting them.

In a raid in Ireland in early 2017, nearly three hundred designer puppies, including Cavachons and labradoodles, were seized at ports because of a clampdown on the illegal selling of puppies. There were multiple raids conducted by customs and the Dublin SPCA. Puppies are bred cheaply in Ireland and sold in the United Kingdom for very high prices. According to Irish news reports these raids sometimes take place as often as once a week. But for every shipment that is stopped, many more make it through. An article published on January 2, 2017 by the Irish website RTÉ.ie states:

> [When discovered,] some of the puppies are shook [unwell] and well under the legal age for export, which is 15 weeks.
>
> Such seizures are the result of thwarted attempts to smuggle dogs and puppies out of Ireland without the correct paperwork or vaccination records.[27]

27 RTÉ, "Clampdown On Illegal Selling of 'Designer Puppies'" (January 4, 2017), https://www.rte.ie/news/special-reports/2016/1224/841065-dogs-dspca/.

The article goes on to say that an earlier seizure of a shipment of ninety-six puppies sent from Dublin into the United Kingdom was valued at "between £70,000 and £75,000."

Just two days after the RTÉ.ie article was published, the *Myanmar Times* published an article by Nick Baker titled, "One reporter, 500 dogs: A visit to the Yangon Animal Shelter."[28] There isn't a large culture of animal adoption in Myanmar, so most dogs live on the streets. The fortunate ones have found a home at the Yangon Animal Shelter, a shelter for dogs that is privately owned and run by Terryl Just, an American expat.

Among the five hundred dogs who call the Yangon Animal Shelter home are paralyzed dogs, poisoned dogs, and ones with a wide variety of problems common to street dogs. One thing that is particularly upsetting is the growing trend of designer dogs being imported to Myanmar from Thailand. People spend about $800 on these dogs, but the dogs end up in the shelter because locals don't really want them. Describing a conversation with Just, Baker wrote:

> She said that most people who adopt from Yangon Animal Shelter are expats or—surprisingly—a few tourists who decide to take a dog back overseas.

> "We have about a dozen dogs around the world," she said, listing a variety of mainly US cities from Los Angeles to San Francisco to Washington, DC. . . .

---

28 Nick Baker, "One Reporter, 500 Dogs: A Visit to the Yangon Animal Shelter," *Myanmar Times* (January 4, 2017), https://www.mmtimes.com/lifestyle/24409-one-reporter-500-dogs-a-visit-to-the-yangon-animal-shelter.html.

"Some countries have very strict quarantine regulations but the US and Canada are very easy," Just said.[29]

Although I object to the American quarantine regulations being classified as "easy," it is clear that we are dealing with a worldwide criminal enterprise in the business of satisfying whims that can operate beyond the capabilities and reach of law enforcement. Instead of heroin, the contraband is designer puppies. The huge sums of money expended on this temporary high, which it is, with many puppies discarded as quickly as they are purchased, is about the same as the drug industry. It is not a coincidence that the businesses overlap when puppies are also used as mules to move drugs.

---

29 Ibid.

CHAPTER 8

SUBPRIME PUPPY MORTGAGE
TO THE RESCUE

The sun, the moon, and the stars would have disappeared long ago . . . had they happened to be within the reach of predatory human hands.

—*HAVELOCK ELLIS*

How are so many people able to afford designer dogs? High-demand purebreds, like the French bulldog, and custom-designed dogs, like a designer -oodles, cost thousands of dollars whether the dog is an original or a knockoff. Designer mutts are often priced higher than the combined cost of the two purebreds used to breed them, and far more than a mutt or a designer dog found in a shelter. In many cases, the dog in the shelter was the dog originally sold for thousands of dollars. (Shelter dogs can go from free to $500 depending on the shelter.) Yet, if you are looking to impress, that fact might mean nothing to you since the animal is still a shelter dog. This snob factor is a key component of explaining this phenomenon, similar to the preference for buying direct from the designer rather than buying the same item from a designer discount store.

When I think of the snob factor, I think of Coco Chanel.

Coco Chanel's marketing theory was never to sell to everyone. She didn't want people who couldn't afford her clothes to wear them. She created a special club of people she wanted dressed in her clothes. Those people paid high prices to be special and have something the average person could not afford. Today there are people who continue to buy clothes from Chanel, and there are people who buy look-alike items to deceive others into thinking they're part of the elite authentic Chanel-wearing world.

In the puppy ownership world, people also want to look like they are part of a wealthy, exclusive society. But in the dog world, it's not always clear if you're buying an authentic name brand or a knockoff, like the Rolex on New York City streets or the Chanel look-alike at a swap meet. A designer dog can't be exposed as fake by a "Made in China" sticker, and it can be easy for a consumer to be misled. Sometimes imposters are sold at the prices of an original.

Where does the average person come up with $5,000 to spend on a designer dog, an authentic or expensive imposter? One way is through pet shops that are often more than happy to connect you with a financing company that is thrilled to support you in leasing a dog. As with subprime mortgages, there is no requirement that you actually be able to afford the dog, a balloon payment, interest rates that exceed usurious levels, and if payments are not made, the dog, now your family member, can be repossessed. Even if you pay the lease installments up to the final balloon payment, you still do not own the dog. Just like a car, it's not yours until you decide to buy it at the end of the lease period. Frequently, the leaser feels hoodwinked by this realization. Many people think they are paying the full cost of the dog by dividing the price into equal installment amounts with some owed interest, and do not realize that it is actually a pricey lease-to-own arrangement. Furthermore, as is the case with a credit default swap tool, the original lender can bundle your debt and assign it to someone else, who can sell it to someone else, and ultimately you don't know who is owed the money or who to talk to in the event of a problem. At this point, the pet store is no longer involved, as they were compensated by the third-party lender. The lender expected the pet store personnel to inform you of all the details. The pet store salesman didn't fully inform you because if he had, you wouldn't have bought the dog in the first place. It's a hard blow to realize your dog is not actually yours.

As is characteristic of subprime lending, only people who can't afford the loan in the first place agree to it and many of them don't understand what they are agreeing to. They often don't understand that the dog belongs to the finance company, not the family. Worse, the person who could not afford a $2,500 dog now might owe $6,000 for the dog. Consumers discovering this—a pet bought for $1,800 really costing almost $4,500, or a $500 pet really costing almost $3,000—have been reported in all fifty states. Of course, since only those people who can't afford the dog (or the medical bills, food, and boarding) opt for this arrangement—they usually also can't afford lawyers to help untangle the problem. That is the nature of a predatory lending scheme; it mostly works on low income and uninformed buyers who can't fight back. It is also the modus operandi of a bully.

On the other hand, there are people who like to lease a car for a few years and then get a new one, and those who can't afford the car but can afford the higher lease payments for the same car. This sort of consumer might like the idea of taking home a dog without having to actually own it. Still, knowledge of the decision being made is important. Informed consent is the key to truly agreeing to contract terms. All car leasers know they are leasing; the same isn't true for all dog leasers.

The finance companies setting up pet leases feel that it is up to the pet shop to disclose all the particulars and that this is not their concern. This is usually accurate if specific disclosure requirements are met. Such requirements are intended to inform the consumer of the terms of the contract, so they can understand the nature of the transaction. If someone who wants to drive a Jaguar can't afford to buy one but can meet the expense of the monthly payments, it's considered to be a win-win deal. The person can drive a high-end car for a few years and feel good doing it. It is all about the disclosure and the informed nature of the decision. The biggest issue with pet leases is that the buyers don't know what they're signing up for.

So, how does this work with a pet? Existing lending laws do not consider their potential application to live animals and therefore lack guidance for handling the issues that can arise. What happens if the pet is sick, bites, dies, or is unsuited to the family? What happens to the pet? What happens to the dog who gets repossessed and can no longer be sold as a cute puppy? That dog may end up in a shelter or sold to another store. There is human grief, and there is dog grief too. Dogs also get depressed, and they can also pine for a lost friend. If you knew the potential results and implications, would the loan be worth it so that you could have an imitation of Hilary Duff's or Leonardo DiCaprio's dog?

Although the pet leasing practice has been in existence for many years, it has only recently drawn public attention, as a result of the designer dog trend. There has been a spike in partic5ipating lenders stepping up to "help" people purchase a designer dog. Except for the fact that consumers don't actually own the dog, it sounds great. Victims crying foul are sprouting up all over the country. Customers who accepted a financing plan without understanding what they signed up for are complaining loudly. This group includes those who decided that they didn't want to keep the dog and yet still owed money, those who wrecked their credit score by defaulting, and those who felt cheated because they weren't told it was a lease and not a purchase.

The situation has become so grave that in 2017 California passed legislation to address it. As of January 1, 2018, these contracts have been deemed "void against public policy" in California. Essentially, this means that even if something is technically legal, the action could still be deemed void. With pets, this means that it has been established that leasing a dog or cat is contrary to society's moral standards. Unlike a car, we are dealing with an animal who can bond to a family and vice versa. Therefore all animals previously labeled as leased are now owned by the buyer. Nevada was the second state to do so and Rhode Island and New York have similar legislation pending. Section 1670.9 of the Civil Code of California states:

> Existing law generally regulates formation and enforcement of contracts, including what constitutes an unlawful contract. Under existing law a contract is unlawful if it is contrary to an express provision of law, contrary to the policy of express law, though not expressly prohibited, or otherwise contrary to good morals. Existing law, the Unruh Act, provides for the regulation of retail installment contracts, as defined. Existing law, the Karnette Rental-Purchase Act, provides for the regulation of rental-purchase agreements, as defined.

> Existing law regulates the sale of dogs and cats in this state, including provisions governing the retail sale of dogs and cats.

This bill would declare a contract entered into on or after January 1, 2018, to transfer ownership of a dog or cat in which ownership is contingent upon the making of payments over a period of time subsequent to the transfer of possession of the dog or cat void as against public policy unless those payments are on an unsecured loan for the purchase of that animal. This bill would also declare a contract entered into, on, or after January 1, 2018, for the lease of a dog or cat void as against public policy. The bill would require that the consumer taking possession of a dog or cat transferred under the terms of one of these contracts be deemed the owner of the dog or cat and be entitled to return all amounts paid under the contract.

SECTION 1.

Section 1670.9 is added to the Civil Code, to read:

1670.9.

(a) (1) Except as provided in paragraph (2), a contract entered into on or after January 1, 2018, to transfer ownership of a dog or cat in which ownership is contingent upon the making of payments over a period of time subsequent to the transfer of possession of the dog or cat is void as against public policy.

(2) Paragraph (1) shall not apply to payments to repay an unsecured loan for the purchase of the dog or cat.

(b) A contract entered into on or after January 1, 2018, for the lease of a dog or cat is void as against public policy.

(c) In addition to any other remedies provided by law, the consumer taking possession of a dog or cat transferred under the terms of a contract described in paragraph (1) of subdivision (a) or in subdivision (b) shall be deemed the owner of the dog or cat and shall be entitled to return all amounts paid under the contract.[30]

Designer dog hysteria is full of predators and bullies. There are those exploiting and preying upon vulnerable puppies, and predatory lenders taking advantage of people who want them. Puppies and leasers are unable to fight back or enlist help.

The cost of cute tripled for just the purchase price—not to mention a potentially wrecked credit score and a visit from a repo guy. You haven't even bought dog food yet.

---

30 California Legislative Information, "AB-1491 Sales of Dogs and Cats: Contracts," Assembly Bill No. 1491 (October 13, 2017), https://leginfo.legislature.ca.gov/faces/billNavClient.xhtml?bill_id=201720180AB1491.

CHAPTER 9

BREEDERS GONE WILD, TO GO WHERE NONE HAVE GONE BEFORE

Our scientific power has outrun our spiritual power. We have guided missiles and misguided men.

—MARTIN LUTHER KING JR.

The urge to create designer mutts, or "Frankendogs," shows no signs of abating. The people want, the people pay, and the people receive. So far, we have been talking about mating dogs with dogs. What about a custom-made pet with one parent that's a dog and another that's a different species?

This concept of a mixed-species pet was embraced by *CatDog*, an animated television series created by Peter Hannan that aired on Nickelodeon from 1998 to 2005. The show was about a cat and a dog who were conjoined twins and shared a torso. The duo, CatDog, has a dog head and a cat head at opposite ends of its dachshund-like body. The episodes chronicle the cat and dog attempting to get along with each other despite often opposing personalities, and also trying to cope in a world in which CatDog doesn't fit, even though he has the skills and talents of both species. My kids and I used to watch this show regularly and we would imagine what animal we would like to be paired with.

Being able to fly, and therefore being half a winged creature, was a popular fantasy in my house. I, however, being an incurable spy and eavesdropper, coveted the power of invisibility. That left me half chameleon. Not exactly my dream come true. Alone in the world as a little femeleon?

The idea of selecting specific traits from different breeds to create a unique third breed is illustrated in the extreme by this conjoined cat and dog. Beyond the fun of the dog's preference for fast food and rock and roll contrasted with the cat's gourmet palate and love of opera, it is the combined specific genetic traits of each that save the day. The cat can leap to great heights and is cunning, while the dog is gullible, loyal, and friendly. Together they are unbeatable as they extricate themselves from all sorts of silly situations.

In 2014, another so-called animal hybrid, this time a real-life one, made headlines: Lion Dog. The animal, a purebred Tibetan mastiff, is one of the most expensive dogs in the world and was reported to have been sold to a wealthy property developer in China for $2 million. Tibetan mastiffs have become the ultimate luxury status symbol in China and are beloved like pandas. At the time of the sale, the dog's breeder Zhang Gengyun was quoted as saying, "They have lion's blood and are top-of-the-range mastiff studs." Following this, some news reports left the impression that there was real lion's blood in the dog. In March 2014 the *Qianjiang Evening News*, a popular Chinese newspaper, explained that "lion's blood" means "very good blood" or Chinese noble blood.[31] The property developer now has two of these dogs and plans to breed them. In August 2013, CNN

---

31 Agence France-Presse, "Dog Sold for Nearly US $2 million at Zhejiang Luxury Pet Fair: Report," *South China Morning Post* (March 19, 2015), https://www.scmp.com/news/china/article/1452381/dog-sold-nearly-us2-million-zhejiang-luxury-pet-fair-report.

reported that a zoo in China actually tried to pass off a Tibetan mastiff as an African lion. Unfortunately, the "lion" began to bark, which blew the entire plan.[32]

Though we are not quite ready for a real-life CatDog, it is fun to ponder the possibilities of genetic engineering. Science is poised to deliver some really interesting things; the idea of combining compatible species is no longer just science fiction. In fact, there are routine medical procedures that currently use animal parts in the human body, such as the use of a pig heart valve in a human heart when the original human heart valve stops functioning properly. The important denominator for medical experts or genetic engineers who mix and match different species or two specimens of the same species is to use scientific evidence to achieve predictable results by adherence to provable theorems and principles. To proceed blindly, as is common in the custom-designed dog industry, yields a lot of costly waste product, specifically, the health and lives of the rejected dogs. In essence, replicability and reaching the same conclusions should occur each time if the math and science involved is correct. This thinking, applied to breeding, could result in breeding practices that help overbred purebred dogs achieve a better ability to breed true, with the caveat that behavior and personality are still not at all predictable.

A Clone Named Blue

In 2011, a man in New Mexico paid $100,000 to clone Old Blue, his dying dog. He sent Old Blue's DNA to a laboratory in South

---

32 CNN Staff, "Chinese Zoo Angers Visitors by Passing Off Hairy Tibetan Mastiff Dog as Lion," CNN (August 16, 2013), https://www.cnn.com/2013/08/16/world/asia/china-zoo-dog-lion/index.html.

Korea while the dog was still alive, and it was stored there until the request came to proceed with the cloning. The new Blue is a genetic replica of Old Blue and appears to be his carbon copy. A couple in England brought samples of their dead dog to South Korea at the end of 2015 and two boxer clones were born.

Cloning dogs is not an exact science and may never succeed in replicating the behavior and personality of the original, but the dogs do look like their originals. But there is physical price paid by the laboratory dogs involved. Eggs have to be extracted from a large number of dogs and cloned embryos must then be implanted into other dogs so that, maybe, a couple of pregnancies might take. Again, a lot of wasted dogs to achieve one desirable dog.

Flash forward to today and anyone can clone their pet for the price of $50,000 for a dog and $25,000 for a cat, much less than the cost of Old Blue. One company providing this service is ViaGen Pets, which bills itself as "America's Pet Cloning and Preservation Experts—Serving Pet Parents Worldwide." Its website states:

> A beloved pet is much like a family member. The unique life-enriching bond, the love and companionship—a truly special pet provides us a unique sense of comfort and life-enriching fulfillment which is nearly impossible to extend beyond your pet's natural life span. Until now.[33]

Let's not go too deeply into this rabbit hole, as it becomes quite technical. However, it is now possible to obtain genetic material from your pet, whether the pet is dead or alive, a significant advancement since the old technology required the pet to be alive. This solves the problem of a sudden and unexpected death

33 Viagen Pets, https://viagenpets.com.

leaving no time to harvest the genetic material. ViaGen further asserts that the resulting pet will be healthy, live a normal life span, and not require special care unique to and derivative of the cloning process. They can guarantee that the gender of the cloned dog or cat will be the same as the original, and that for the most part they will look alike, but there are no guarantees regarding the personality or behavior of the cloned animal. The ViaGen website states:

> Cats and dogs delivered by cloning have the same genes as their donor pets and will be the closest match possible to the donor. This is best described as identical twins born at a later date. The new puppies and kittens will be the same sex as the donor, but just as it is in nature, may have slight phenotypic differences, such as different markings due to natural epigenetic factors. The environment does interact with genetics to impact many traits such as personality and behavior.[34]

Cloning pets has been increasing over the years, but slowly. Recently, however, celebrities have received attention for doing it, and this makes me fear that a trend may start. Barbra Streisand cloned her dog Sammie, who died in 2017. She used ViaGen and ended up with four clones; she kept one and gave three to family and friends. Diane von Furstenberg paid a Korean company $100,000 to clone her Jack Russell terrier Shannon and received two cloned Shannons back. She named them Deena and Evita.

I understand the attachment and love one feels for a dog. Yet, when I think about the millions of dogs in the United States who are killed for no other reason than insufficient time to wait for a home, or all the wonderful dogs who we find in shelters, I am

---

34 Ibid.

concerned about the idea of mass cloning. So many already-born pets need homes. When I reflect back on each time I thought my dog was the most unique and wonderful dog in the universe, it makes me accept death as a way of life and I celebrate the differences in each of our pets.

The actress Jo Anne Worley approached the loss of her dog differently. When her constant companion, a Yorkie named Harmony, passed away, she asked me to find her a dog that resembled Harmony. This was enlightened as she sought the clone effect, but wanted to adopt from a shelter rather than purchase from a breeder or clone company. I actually found an exact replica of Harmony in my shelter, but when she saw the dog, rather than feeling better, she fell apart and realized that seeing an image of her dog aggravated her grief even more. Instead, she fell in love with and adopted Cupid, a tan Chihuahua from my shelter. People handle grief differently. Cupid and Jo Anne have become inseparable, and instead of ordering a new designer dog, Jo Anne dresses Cupid in the best designer dog wear augmented by the best designer accessories. You too could find your dog's "clone" in a shelter. You want her or you might decide on another dog, as Jo Anne did, but either way is better than having one made.

When I consider that people are willing to spend $25,000, $50,000, $100,000, or more on a cloned pet, I know it's because they really love their pets and I understand that they don't want to let them go. But cloning only provides the physical replica, not the personality, behavior, and soul of the lost pet. I would ask them to, instead, adopt one from a shelter and donate the remaining funds to help other pets find health, happiness, and a home. I can also guarantee that even a $100,000 dog will drink from the toilet bowl and eat cat poop.

To Russia With Love

What if you could take genetic material for specific traits, looks, and skills and make your perfect custom-designed dog? You could mix breeds, change appearances, and create a dog the way you would build a teddy bear at the mall. This is not just a crazy science idea from a literature major with an overactive imagination.

At the close of 2016, Vladimir Putin received the ultimate cloned designer war dog. Three genetically modified Belgian Malinois puppies were engineered to be strong and highly capable of sniffing out drugs and explosives. They were designed by Dr. Hwang Woo-suk at the Sooam Biotech Research Foundation in Seoul, South Korea, the same scientist who cloned Old Blue—and who, by the way, is working on restoring the wooly mammoth. Costing about $100,000 apiece, the dogs were bred from the genetic material of the best sniffer dogs in Korea. Unfortunately, these supposed "super war dogs" failed basic obedience and skill tests, did not learn Russian commands (they understood Korean), and hated the Siberian cold. Putin scrapped the entire program and assigned the dogs to guard prisoners at Forced Labor Camp #1 in Yakutsk. I am not holding my breath for the return of the woolly mammoth.

If you are thinking this seems very *Jurassic Park*, you're not wrong. As in *Jurassic Park* when the proprietor of the dinosaur theme park assures the guests that the dinosaurs can't reproduce due to genetics, Dr. Ian Malcolm, played by Jeff Goldblum, nervously says, "Life . . . uh . . . finds a way."

If you think this type of engineering is relegated to science fiction and Russian war dogs, you would be incorrect. We already know that designer dogs are created to minimize allergies, to serve as war dogs, to fight, and to be smaller or funnier looking than other dogs—and that people's imaginations are limitless. If you can conceive it, there's someone out there who's willing to try to attain it. At present, breeding two dogs to get the dog that you have in mind is a game of luck. You may not get the looks, coat, or state of health that you want, but what if, through cloning technology, you could manipulate DNA and either breed in or breed out a desired result, cloning for good or evil, if you will?

Can You Breed a Dog to Match My Eyes?

Some genome-editing tools come in relatively inexpensive kits and their availability and accessibility can allow anyone, scientist or backyard breeder, to influence the DNA of a dog. In the right hands, they may make it possible to unbreed some of the damage caused by inbreeding and crossbreeding, or natural genetic inheritance. In other words, you could take a golden retriever line that is prone to cancer and change that prognosis and the breed's destiny.

In a March 2, 2017 article published by the *San Diego Union-Tribune*, Chris Reed asks this question in the title of his piece: "From cloning Dolly the sheep in a lab to gene editing dogs in a shed: progress?"[35] In his article, Reed mentions David Ishee, a Mississippi dog breeder with a GED and no advanced science degree or credentials, who uses a kit and DNA that he orders online to selectively breed dogs. He tries to undo the damage caused by unfettered inbreeding and poor breeding by editing out bad genes and creating more genetic diversity. Reed writes that Ishee would like to edit the gene that causes a fatal bladder disease in dalmatians.

---

35 Chris Reed, "From Cloning Dolly the Sheep in a Lab to Gene Editing Dogs in a Shed: Progress?" *The San Diego Union Tribune* (March 2, 2017), http://www.sandiegouniontribune.com/opinion/sd-cloning-gene-editing-designer-babies-20170301-story.html.

In a February 26, 2017 posting by Vanessa Bates Ramirez on the website Singularity Hub, Ramirez asks the question, "Would you want a dog that was genetically engineered to be healthier?"[36] The answer is, why would you not, if you could cure diseases and ease a lifetime of discomfort for your dog? Ramirez also wrote about David Ishee and his ability to order synthetic DNA online. She quotes him as follows:

> The biggest thing here is the collapsing price of DNA sequencing and synthesis. You can order synthetic DNA for about nine cents a base pair. When I ordered my construct a year and a half ago, I paid 23 cents a base pair. Six years before that it would've been $1.30 a base pair. When it gets down to pennies, people will be able to do much more complex things.

Who Knew Amazon Had a DNA Section?

The concept of anyone being able to edit genes from their home, with supplies ordered online, can be a game changer for designer dogs. A fully sequenced dog genome and genetic tools that can reveal gene mutations and assist ethical breeders to better predict congenital diseases and a few physical traits already exist. Of course, this field is fraught with all sorts of ethical considerations and dangers depending on those who are doing the editing. One person may want to cure heart disease while another might edit for appearance only.

Sharing the power to use or abuse scientific knowledge will come down to a consensus on ethical considerations. Just because

36 Vanessa Bates Ramirez, "Would You Want a Dog That Was Genetically Engineered to Be Healthier?" SingularityHub (February 26, 2017), https://singularityhub.com/2017/02/26/would-you-want-a-dog-that-was-genetically-engineered-to-be-healthier/#sm.000005243z4tzenwtyo13bzh5basj.

we *can* do something, *should* we do it? This question can be answered in many ways. Yes, because it can help make perfect specimens. No, because we could end up living in a futuristic world like the one depicted in the film *Gattaca*, in which natural breeding of humans is prohibited to avoid surprise traits and the government is in charge of determining what traits to breed.

The FDA is currently imposing strict regulations and restrictions on manipulating the genes of mammals, since this type of science could be applied to human babies as well as other creatures. They've decided that it's okay to edit dog genomes as long as it is in the interest of the dog's health, saying in essence that an edited dog genome is like a veterinary drug. (As such, the genome is subjected to all regulations and required testing of a new veterinary drug.) Clearly, it is the intention of the government to control this area. If used intelligently, it could save breeds from extinction, fix what has happened to breeds that has given them a bad reputation, and cure diseases. If not, it will make the existing situation exponentially worse, as uneducated breeders will continue to play god, with new toys at their disposal, and command even higher prices for the results of the experiments. Below are excerpts from the FDA's website that apply to animals and veterinary medicine:

> Recent scientific advances now make it possible to more efficiently and precisely alter the genome of plants, animals, and microorganisms to produce desired traits. These genome editing technologies are relatively easy to use and can be applied broadly across the medical, food and environmental sectors, with potentially profound beneficial effects on human and animal health. However, there are also potential risks ranging from how the technology affects individual genomes to its

potential environmental and ecosystem impacts. Additionally, genome editing has raised fundamental ethical questions about human and animal life. . . .

When *animals are* produced using genome editing, FDA has determined that, unless otherwise excluded, the portion of an animal's genome that has been intentionally altered, whether mediated by rDNA or modern genome editing technologies, is a drug because it is intended to alter the structure or function of the animal and, thus, subject to regulation under our provisions for new animal drugs. We have updated our existing guidance for genetically engineered animals to include genome editing within its scope, and are issuing it in draft form for public comment. We are also seeking input on whether certain types of genome editing in animals pose low or no significant risk, and we may modify our regulatory approach based on this input.[37]

The FDA neither controls what goes on in backyard sheds nor in labs in other countries. One has to wonder how serious a deterrent their regulations would be to a non-federally funded scientist—or a self-proclaimed scientist. Given the slack enforcement of other federal laws, I wonder what compliance with these directives, if any, would look like.

I predict that we will be seeing a lot more people playing with genome editing. If anyone reading this can make genetic corrections on existing people, please put me down to grow six more inches with thick straight hair and no cavities.

---

37 Robert M. Califf, MD, and Ritu Nalubola, PhD, "FDA's Science-Based Approach to Genome Edited Products," FDA Voice (January 18, 2017), U.S. Department of Health and Human Services, U.S. Food and Drug Administration, https://blogs.fda.gov.

CHAPTER 10

MYTHS AND MAGICAL THINKING

If fifty million people say a foolish thing, it is still a foolish thing.

—ANATOLE FRANCE

If puppy mills and unethical breeding practices don't bother you, trafficking and transport issues don't concern you, and the treatment of animals in minimill-style pet shops is okay with you, perhaps it's because you have the healthy, behaviorally sound, brilliant, hypoallergenic, unique, elite, high-end, adorable, trendy designer dog that you always wanted, that was worth the thousands of dollars you paid for it. If this is the case, it's possible that you will be happy and your dog will be loved for his or her entire life. If this is not the case, you may already be or end up as one of the alarmingly high number of people who have relinquished their dog to a shelter or the streets, or has been forced to euthanize the dog after expensive and exhaustive medical efforts failed. You might also have discovered that your expectation to spend many wonderful years with your dog turned out to be forty-eight hours or a few tormented years. Was the decision to buy this dog based on myths and magical thinking, or on fact? Were you shocked, or did you know what you were getting into and plan accordingly? As Rachel Maddow would say, "Let's pay a visit to Debunktion Junction."

Myth: A Registered Facility Won't Be a Puppy Mill

Even if a facility or breeder is registered with the American Kennel Club (AKC) or USDA, there is no guarantee that its dogs are healthy and were bred by a responsible, ethical, and credentialed breeder. AKC is painfully aware that they have not always been interested in looking into their high-volume breeder members. (AKC doesn't refer to them as "puppy mills.") In a report from November 12, 2002, titled "High Volume Breeders Committee Report to the American Kennel Club Board of Directors," they acknowledged that they have traditionally been paper checkers and fee collectors, and did not check the conditions of animals.

> AKC had no program of kennel inspections. When it instituted such programs they were limited to verifying registration records and dog identification. Prior to 1996 AKC had no sanctions related to the conditions under which AKC registered dogs were raised or the care they received. In those times, it was AKC's position that the conditions of a kennel were solely the province of USDA and local humane societies and law enforcement bodies.

> When AKC inspectors were dispatched to kennels, they were expected and even instructed to ignore the conditions they found there, and to audit only the records and identification of the dogs, not their health or the conditions in which they were kept.[38]

38 The American Kennel Club, "High Volume Breeders Committee Report to The American Kennel Club Board of Directors" (November 12, 2002).

In 1996, AKC adopted requirements making adherence to minimum standards a prerequisite for listing on their registry, and said they would begin active inspections to ensure that minimum standards were met. They made it clear that they only register the dogs, not the place, and only report violations to law enforcement. They also now offer a variety of free breeder education classes. However, exposés and complaints from people who purchased AKC-registered pets would suggest that their plan is not working as hoped. For example, in 2013, NBC's *Today* show ran a story that featured an AKC-registered establishment even though AKC claimed they wouldn't register establishments. The place was a puppy mill, and the dogs inside were being raised in wretched and inhumane conditions. Purchasers were interviewed for the show, and they said they received very sick dogs from AKC-registered breeders. There was a hint of conflict of interest that raised the question of whether the payment of registration fees by the breeders made it impossible for AKC to conduct an honest inspection.

It is best not to assume that a membership club or even the USDA (which also collects registration fees) has eyes everywhere and that when they see something wrong they will do the right thing, even if it is not in their interest to do so. It is far better is to research local reputable breeders where you can visit, ask questions, bring a veterinarian with you, and verify documentation, rather than just rely on a membership registration or order a dog from the internet. You may have to wait a little longer to get your pet, but you will be glad you did.

Myth: Designer Dogs Are Hypoallergenic

There is no such thing as a truly hypoallergenic dog and yet millions of allergy sufferers search for one that is. There are dogs whose coat dander and saliva may be less bothersome than that of other dogs, but dogs also carry pollen, dust, and mold in their coats, which may also produce a reaction, and unless a person has tested the hairs and saliva against his specific allergy condition, one can't be sure that there will be any mitigation of symptoms. A study published by the *American Journal of Rhinology and Allergy* in July 2011 suggests that there may be no difference between allergens present in a "hypoallergenic dog" and a "non-hypoallergenic dog."[39] For the study, researchers collected dust samples from the floors of nurseries and analyzed them for the presence of dog allergens. They found no significant difference based on whether the dog was considered to be a hypoallergenic breed or not. Summing up the findings in an article for the website MedPage Today ("Hypoallergenic Dogs May Not Protect Against Allergies," July 12, 2011), Todd Neale wrote:

Homes with a hypoallergenic dog were no less likely to have detectable levels of dog allergen or to have lower average levels of allergen than homes with a non-hypoallergenic breed, according to Charlotte Nicholas, MPH, of Henry Ford Health System in Detroit, and colleagues. . . .

In homes where the dog was allowed in the baby's bedroom, allergen levels tended to be lower with hypoallergenic dogs, and in homes where the dog was not allowed in the bedroom, levels

39 Charlotte Nicholas, Ganesa Wegienka, Suzanne Havstad, Edward Zoratti, Dennis Ownby, and Christine Johnson, "Dog Allergen Levels In Homes with Hypoallergenic Compared with Nonhypoallergenic Dogs," *American Journal of Rhinology and Allergy* 25(4): 252-6 (July 2011), https://www.ncbi.nlm.nih.gov/pmc/articles/PMC3680143/.

tended to be higher with hypoallergenic dogs, although none of the differences reached statistical significance.[40]

As this is an important subject and one responsible for creating an abundance of "oodle" mixes, here is one more statement on the subject to emphasize the point. This is from an article titled "Pet Allergy" on the website of the American Academy of Allergy Asthma & Immunology:

Almost 62% of US households have pets, and more than 161 million of these pets are cats and dogs. Unfortunately, millions of pet owners have an allergy to their animals.

The proteins found in a pet's dander, skin flakes, saliva and urine can cause an allergic reaction or aggravate asthma symptoms in some people. Also, pet hair or fur can collect pollen, mold spores and other outdoor allergens. . . .

Contrary to popular opinion, there are no truly "hypoallergenic breeds" of dogs or cats. Allergic dander in cats and dogs is not affected by length of hair or fur, nor by the amount of shedding.[41]

There is no good reason to pay a huge price for a miracle dog, patronize puppy mills that breed such dogs, or believe in representations from snake oil salesmen. But there is a need to make sure the allergy sufferer can tolerate a specific pet through trial and error, using hair and saliva samples. Washing a pet often,

---

40 Todd Neale, "'Hypoallergenic' Dogs May Not Protect Against Allergies," MedPage Today (July 12, 2011), https://www.medpagetoday.com/allergyimmunology/allergy/27503.

41 American Academy of Allergy, Asthma, and Immunology, "Pet Allergy," https://www.aaaai.org/conditions-and-treatments/allergies/pet-allergy.

getting rid of carpets, using HEPA air filters, and vacuuming frequently might also help the allergy sufferer.

Myth: Mixing Breeds Results in the Best Traits Being Passed Forward

Will a custom-designed dog result in a healthier dog? Rarely. Perhaps with a responsible breeder who is crossbreeding first generation dogs who were also responsibly bred, there is a chance that there might be some "hybrid vigor," which is essentially a beneficial enhancement achieved through the diversity of crossbreeding. Most custom-designed dogs are not bred with care in the selection of the parents, or extreme care to avoid bad recessive genetic traits. By carelessly breeding and rebreeding overbred and inbred dogs together, illnesses and defects are virtually guaranteed in the offspring. For example, both Labradors and poodles tend to suffer from hip dysplasia, epilepsy, eye problems, and Addison's disease. Therefore, when you breed a poodle with a Labrador, you are just about guaranteeing that the labradoodle will have those issues, and more, depending on how inbred the parents were. The same is true if the breed is between two labradoodles.

If the breeder is working with a tiny gene pool as a result of inbreeding, the worst traits will come through. If a breeder is mixing two dogs to create a custom dog, a nightmare can occur. We have seen instances where dogs who are clearly the product of mixing two breeds for a specific feature end up in the shelter. Some people think that dogs like pugs, Pekingese, and French bulldogs have comically bulging eyes and that crossbreeding them can create a hilarious, bug-eyed mutt. In reality, the result often causes a condition in which the eyes keep falling out of their sockets. The dogs end up in shelters because either the buyer freaked out the first time the eyes came out, or the seller knew the dog could not be sold easily without remedial (if possible) surgery. Even with treatment, the dog may be blind. Not so funny anymore.

Similarly, a dog whose tongue is always out, sometimes sideways, never fails to get a chuckle. We all laugh at a funny face. Hanging tongue syndrome, however, is no laughing matter and is increasingly present due to the reckless breeding practices inherent in creating custom-made dogs. The syndrome is caused by a neurological defect that renders the dog incapable of retracting his or her tongue. Keeping the tongue out of the mouth results in it drying out, cracking, and succumbing to painful bacterial infections. Furthermore, since dogs don't sweat, they rely on panting to expel humidity from the tongue in order to cool down. In the case of flat-faced dogs such as bulldogs, pugs, shih tzu, and Pekingese, who constantly overheat, a dry tongue, though comical looking to some people, can be deadly.

Myth: Teacups, Micro, and Mini Dogs Are Smaller Versions of Large Dogs and Just as Healthy

Nature-made teacups, as you well know by now, are a myth. These man-made dogs were given the label of "teacup" as a marketing ploy. The image of a teeny dog peering over the rim of a teacup is a compelling one, even to me, and I know better. It's natural to find these dogs cute. But teacup is not a charming designation; dogs given this label are an assortment of birth anomalies and defects wrapped in a teensy fur coat. Most come with horrible health problems, are subject to panic attacks, are prone to accidents, and are abandoned quickly when the owner realizes the cost of keeping them alive. Many are bred in awful, unregulated puppy mills in South Korea and elsewhere, and suffer horribly at the hands of those competing to make each tinier than the last. Teacup mothers often die in childbirth, as the process is too strenuous for such an infirm dog with so many medical issues. For breeders, the price of the offspring more than compensates for that loss.

Teacups possess the usual puppy mill litany of illnesses, such as hypoglycemia, heart problems, liver shunts, and respiratory issues, in addition to trisomy (a chromosome disorder similar to Down syndrome), bone abnormalities like cleft paws, skull soft spots, and fragile limbs, to name just a few. Despite this, deformities and all, they are often still irresistibly cute. But this kind of cuteness kills them. Celebrities like Madonna, Paris Hilton, Adele, and Mickey Rourke can afford to locate ethical breeders and pay for the dogs' medical care, but most copycats cannot, and end up discarding these sad and sick little dogs. Here's the kicker: sometimes the unscrupulous will sell a premature puppy as a teacup for the same high price. The dog will grow large, but no refund will be given.

Myth: A Designer Dog Will Be Better Behaved

A dog's behavior, like its personality, cannot be guaranteed, even if the source is an ethical, expert breeder or the dog is a clone of a "perfect" dog. Since behavioral problems come with poorly bred dogs, it is more likely that an expensive designer dog will have them.

Dogs with severe behavioral problems are the most difficult to rehome, adopt, or retain, and often pay the ultimate price with their lives.

Signs Your Dog Was Poorly Bred

Because we are discussing the myths and magical thinking surrounding designer dogs, whether purebred or custom-made, it behooves us to review the most common indicators that your dog was not purebred, but rather poorly bred. These ailments may be present at the time of purchase or may present themselves months or years later.

Parasitic infections from intestinal parasites and worms such as giardia and coccidia. The dog will suffer, much like we do with an intestinal flu or giardia, and could emit mucous diarrhea all over you and your house. If not treated, the dog will fail to thrive and may even die from depletion and dehydration.

Distemper, a highly contagious disease caused by a virus that can attack the intestinal, respiratory, and nervous systems of the dog.

Parvovirus, a contagious virus that presents like distemper but can also damage a heart muscle in young and unborn puppies.

Rabies, a virus that attacks the nervous system and can spread to humans through a bite. It is a particular concern with dogs coming from countries that have not controlled the disease.

Respiratory problems such as bronchial infections, pneumonia, and kennel cough, which present with coughing, sneezing, wheezing, congested lungs, nasal and eye discharges, and trouble breathing.

External parasites, such as mange and scabies, which present as skin rash, crusting, itching, and discomfort. This is contagious to humans and requires rigorous cleaning of the animal's bedding, hairbrushes, and clothing, as well as your own bed and furniture.

A host of other problems, such as ear mites, ear infections, ringworm, fleas, tics, urinary problems, hypoglycemia, and bladder issues.

Weeks, months, or even a year after purchase, the dog may start to exhibit the presence of a congenital or inherited illness, usually the direct result of inbreeding and the use of an insufficiently diverse gene pool. Abnormalities like these can happen in the best-bred dogs when a recessive or ancestral gene pops up, but dogs coming from puppy mills or backyard breeders are most likely to manifest one or more of these problems. These include, but are not limited to:

Seizures or other neurological problems, heart defects, hernias, deafness, blindness, cherry eye, cataracts, and spinal disorders.

A liver disease known as liver shunt, a condition due to a birth defect in which the blood bypasses the liver. The liver cannot survive this and neither can the dog.

Other conditions such as luxating patella (trick knees), collapsed tracheas, Cushing's disease, cleft palate, and excessive sensitivity to heat are very common in inbred small-breed dogs.

All of these conditions require expensive veterinary care, often surgery, and constant care throughout the dog's life. Left untreated the conditions can be fatal. They can also spread to other dogs. As these sick dogs are frequently abandoned or dumped at shelters by unhappy buyers, shelters are saddled with the burden of determining whether the dog can be rehabilitated and enjoy a good quality of life in the right home, will be in constant pain even with medical intervention, or is irremediably suffering. Keeping a dog on strong painkillers for his or her entire life is not usually a sound medical solution.

Behavioral and personality traits are also affected by inbreeding and negligent breeding. We see behavioral issues ranging from obsessive-compulsive disorder, which in dogs often manifests as running in circles all day, to aggression, fear, and biting. All affect the quality of the dog's life and the likelihood of the dog finding a home. Dogs with behavioral issues are the most difficult to place. If the behavior is biting, placing the dog with a family is usually impossible for liability concerns and because no dog should be muzzled all day.

The cost of cute includes purchase costs, medical care, prescription diets, a life of pain, poor credit scores, muzzles, doggie Prozac, homelessness, and possibly the ultimate price.

How to Spot a Reputable Breeder

Despite what you may think, I have friends who have purchased dogs from breeders. They get nervous when I visit as I tend to growl at them for not adopting a shelter dog. It is not uncommon for me to hear "Hide the whippet" when I ring the bell.

I once left a party with someone's designer dog since the dog was not being treated properly, not because the owners were cruel, but rather because the dog had so many digestive and skin issues that she made a mess in the house, she smelled, and no one wanted to go near her. As a result, she was usually left in a cage in the yard. She was a very high maintenance dog with a very expensive veterinarian. I brought her to spcaLA, where we fixed her up and found her a new home. I subsequently learned that the new family adored her and that she even got a job in a commercial with her actor dad. Suffice it to say, I was not invited back to the house I took her from.

If you must have a designer dog, then you must find a reputable breeder to purchase from. A reputable breeder will not sell dogs online. They may advertise that way, but they would insist on making the sale in person. They evaluate the suitability of the buyer as much as you evaluate the suitability of the dog. A reputable breeder only deals in one or two breeds, ones they are expert in handling, rather than multiple breeds. There will not be multiple litters, as protecting the health of the mother is paramount. In fact, you might be told when a litter is expected and asked to wait until then.

There will be no teacups.

A reputable breeder is not ashamed to have you look around the breeding property and won't have any problem if you bring people with you. Insisting upon a parking lot rendezvous is a flashing red warning sign that something is amiss. Check with the Better Business Bureau, the Department of Consumer Affairs, Yelp, Google, and past and present customers for references. Expect to sign a comprehensive contract in which you agree to take proper care of the dog and the breeder agrees to abide by the relevant laws governing refunds, eligible medical reimbursements, and replacement policies. Finally, some breeds, such as English bulldogs, are high maintenance and will always have medical issues, no matter how well-bred they are.

Research the breed before choosing it, and make sure you select a veterinarian who knows the breed well. Then determine whether you can afford the upkeep in addition to the purchase price. Most ethical breeders voluntarily abide by state laws to sterilize the dog before turning him or her over. Clearly, this might stop an existing or aspiring backyard breeder from breeding at home. But if the purpose of buying the pet is to add a new family member, this requirement should not pose a barrier to the sale, and in your area it may simply be the law.

CHAPTER 11

WHERE, OH, WHERE, HAS MY DESIGNER DOG GONE?

The trouble with a kitten is that eventually it becomes a cat.

—OGDEN NASH

As I look at this from a distance, I see a bizarre pattern. Instead of the trope being "the more we make, the less we need; the less we need, the less we make," the opposite is happening. Puppies are being churned out in breathtakingly high numbers, and despite people giving them up we still make more dogs. Is it all impulse and boredom? Are people trying another dog since the first one didn't work out? Or has the diminishment of other breeds due to the custom-designed dog trend resulted in fewer choices, so there is a nudge back to the designer mix, of which there is a surplus?

The Doritos Theory

Even though everyone has to have one, why are designer dogs, both purebred and custom-designed, being abandoned in record numbers in the United States and around the world, without a corresponding decline in production? One reason could be that we live in what I call a Doritos world, where greed, disposability, and carelessness toward things, even living creatures, rules the day.

Almost sixty years after the signing of the Animal Welfare Act (AWA) into law, the situation is actually worse. Most puppies still come from puppy mills, where the breeding practices have not improved since 1966. Today there is more experimentation with custom-designed breeds, and the teacup trend is hotter than ever. Unlike in 1966 when publicity about puppy mill conditions and stolen dogs became a catalyst for change that resulted in the AWA, today's exposés seem to simply add static to the noise, with no signal penetrating. The worldwide designer dog mania seems to have become the shiny object that is distracting us from the moral issues of animal cruelty and pet trafficking. These dogs are more in demand than ever before, and yet they are being mercilessly killed, discarded, and abandoned.

Sometimes people simply lose interest in their dogs or get bored and give them up after the novelty wears off. Other times, they give them up to make room for the next best thing. The main reason that dogs are abandoned is their owner's horrifying realization of how much care these sick puppies need. The must-have allure masks the expense and time commitment. Healthy puppies are high maintenance; ones suffering from disease or congenital illness entail serious commitment and expense.

Shelters are inundated with the detritus of every new trend. Many dogs are brought in from the street and are too far gone to treat. Some are sold cheaply to a swap meet vendor who does not provide the necessary medical care, so the dog may end up resold again or dumped on the street or at a shelter. People give up wanting to care for them or never wanted to in the first place.

When asked why a dog is being surrendered to a shelter, the responses are extraordinary. "The dog is sick and shouldn't be at the price I paid." "I didn't expect to have to feed or clean up after a dog this small." "My children can't play with the dog without it breaking a leg." "The dog smells and I don't know why and don't care." "I don't actually like dogs." "I stole the dog and need to get rid of it." "I'm allergic." "The dog runs in circles all day." "Apparently my 'no pets' lease clause applies to teacups." "The dog is stupid."

In England, pugs, dachshunds, and French bulldogs are being found on the streets in critical condition and turned into the Royal Society for the Prevention of Cruelty to Animals (RSPCA), Dogs Trust, and other charities. The shelters there are inundated with puggles, lurchers (a crossbred dog that is part greyhound), and other designer puppies. Some were Instagram infatuations; all were discarded. The working theory used to explain this extreme dumping phenomenon is that the ability to make impulse purchases over the internet and aping celebrities brought these dogs to people who did not really want them and are in no position to properly care for them. The dogs seem to have no intrinsic value of their own as sentient creatures. Unfortunately, celebrities also junk their pets, behavior that is mimicked and provides justification for those who do the same. Justin Bieber, Kim Kardashian and Ellen DeGeneres have all handed their pets off to someone else. Easy to get, easy to give up.

This feeds into my Doritos theory. Doritos commercials say, "Crunch all you want, we'll make more." What is the message of that? Don't cherish what you have because there will always be more? Keep eating as much as you can? Eat them all? Greed is good? Does it intend to validate the carelessness of a culture of wastefulness and disposability? This permission to have as much as you want doesn't foster a mindful attitude. It's also not true. If it were, there would be no endangered or extinct species. It's not true that people can hunt as much as they want because there's always more. There is a huge, potentially irreversible cost to embracing this attitude. The puppy mill breeders did not create the Doritos world, but they figured out how to feed it. Uncontrolled eating of the chips will lead to weight gain, salt retention, a financial windfall for the company, and maybe a lack of market diversity. Bad things happen when *anything* is uncontrolled. We should not consume just because we can. In the dog world, we haven't found homes for the existing animals in shelters. We don't need so many more; we are careless with what we already have.

The flotsam and jetsam of this industry is everywhere. There are broken dogs and shattered hearts wherever there's a puppy mill. Two purebreds are required to make one crossbreed, increasing the number of misused animals. Go ahead, we'll make more. They cost thousands of dollars, often more than the cost of the two purebreds combined, with thousands more in medical expenses, payments that are necessary whether you own or lease the dog.

Puppy Background Checks

Before you select a dog, you need to know what you're getting into. Tiny teacup puppies are extremely fragile and injure easily through handling and manhandling. It's not uncommon for their bones to be broken while playing or even from simple movements. Teacups often suffer from respiratory distress and heart problems while stuffed in a pocket or purse, either because they panic or because they can't breathe.

English bulldogs are squished-nose, flat-faced breeds, and are extremely sensitive to heat and susceptible to heat stroke. Many new parents don't fully understand this and can overheat the dog just by playing ball with him. You have to be careful when they horse around, enter a hot car, or accompany the family to a little league game on a hot day. This sensitivity is also why this dog has breeding issues, as natural breeding or labor and delivery can be too much for the dog. This was sadly illustrated in a high-profile case at Louisiana Tech University.

In 1899, students at Louisiana Tech found an old stray English bulldog and persuaded the owner of their boarding house to let the dog stay. That same night a fire broke out in the boarding house and the bulldog ran through the house alerting the students until all were safely outside. Unfortunately, the dog did not survive the fire. The university commemorated this noble act by declaring the English bulldog the school mascot and has employed an English bulldog in that capacity ever since.

On a hot day in August 2012, an employee let the then current mascot, named Tech XX, outside to use the bathroom and forgot to let him back in quickly. Tech XX succumbed to heatstroke. To distract authorities from his action, the employee reported the dog missing. Students searched for the dog and a reward for his safe return was advertised until the truth was discovered. The dog's caretaker then issued the following statement:

> Regretfully, I learned this morning that through negligence of an employee, Tech XX was left outside too long on Sunday evening and passed away from a heat stroke.[42]

---

42 Jason Samenow, "Tales of Dogs in Hot Weather: Heart-Wrenching and Heart-Warming," *The Washington Post* (August 14, 2012), https://www.washingtonpost.com/blogs/ capital-weather-gang/post/tales-of-dogs-in-toasty-weather-heart-wrenching-and-heart-warming/2012/08/14/2a43174a-e24d-11e1-98e7-89d659f9c106_blog. html?utm_term=.463bd179b649.

It was later discovered that Tech XIX (Tech XX's predecessor) was retired in 2007 due to health issues related to heatstroke. This may explain the motive for the cover up; the employee should have known to be more careful. In 2016, Tech XXI, who served for three years, was retired at the recommendation of his veterinarian and the USDA. In January 2018, Tech XXII, another English bulldog became the mascot.

This is a good time to think about whether the bulldog of 1899 could be commemorated in a different way. The English bulldog is an inappropriate choice for a dog that will be frolicking around in the heat and humidity of Louisiana. Given the school's track record, as well the original 1899 event, it seems clear that the mascot job is not without danger. Perhaps a person in a climate-controlled bulldog costume would be a smarter choice. Human negligence notwithstanding, exposing dogs to such heat seems an odd way to pay homage to a hero.

Often, a designer dog will end up in a cage or tied to a tree all day. Because of the medical issues, fragility, and neediness, people lose interest. If they feel obligated to keep the dog because of the high price they paid, the dog often ends up kept but ignored. The dog will then either bark incessantly for attention out of boredom or quietly slip into depression. Neither is consistent with a good quality of life.

In many cases, the dog will be sold privately or through a site like Craigslist in an effort to recoup some of the costs. Owners also often attempt to sell their dogs to a pet shops, swap meet vendors, or any willing dealers. Afraid of losing the sale, the owner typically does not disclose all of the dog's medical and behavioral issues, and then the process is repeated with the dog given to yet another owner, turned into a shelter, or euthanized. Other times, the dogs are abandoned in the streets and left to fend for themselves.

This unloading of designer dogs is happening in astronomical numbers. The excitement of a point-and-click purchase of a cute dog seen on social media deteriorates rapidly, from the whimsical joy of having a new toy to the reality of routine care requirements and being saddled with unforeseen medical problems and sicknesses.

Some designer dogs are mishandled, and others are shunned. Dog breeds that became very popular through a film or celebrity sighting can begin to develop a bad reputation for illness or bad behavior, due to ignorant and misguided breeding practices to fulfill the demand. When word gets out, people begin to shun the breed. This happened to collies, Dobermans, dalmatians, cocker spaniels, and Old English sheepdogs, to name a few. A pause in breeding combined with subsequent careful breeding was needed, and this helped preserve these breeds and bring them back a little.

Some breeds, because of deranged breeding practices, can be banned and eradicated. Remember the cautionary tale of the pit bull. The breed is now subject to, and a victim of, breed-specific legislation, some of which actually ban the dog. The pit bull, one of the original designer mutts, is being actively pushed toward extinction.

In the spring of 2012, Chris Brown, rap star, known domestic violence offender, and all around tough guy, was breeding and selling pit bull puppies. On Twitter, he hawked eight-week-old pit bull puppies for $1,000 apiece. Brown crafted the message that by owning a pit bull, you could become a bad boy with a bad dog. This was in sharp contrast with the stereotypical, feminized image of a designer dog and designer dog owner, a party girl with a little puffball of a puppy in her arm. The message was that brutes everywhere need a tough dog as an accessory—after all, men accessorize too. What does Brown know about genetics and responsible breeding? Who was he selling to over Twitter? We don't know, but those who wanted a Chris Brown dog, clearly a name-brand dog, bought them.

We also know that animal shelters across the country are lousy with unwanted pit bull dogs. Adopters spurn them because they fear that the dogs have a violent disposition, as they are associated with dogfighting, gang violence, and other crimes. Many home insurance companies won't insure homeowners with pit bulls as pets—or any other bully breed for that matter. The breed is completely banned in some cities, and some animal shelters won't adopt them out as a matter of policy.

In 2011, pitcher Mark Buehrle signed with the Miami Marlins, but could not live near the Miami stadium because one of his four dogs, Slater, is a pit bull and Miami-Dade County has a pit bull ban. Buehrle, a known dog lover who publicly criticized the dogfighting activities of NFL quarterback Michael Vick, reportedly told the *Miami Herald*, "It's kind of ridiculous that because of the way a dog looks, people will ban it. Every kind of dog has good and bad, and that depends on the handlers. If you leave a dog outside all the time, it'll be crazy. Slater would never do anything harmful."[43] Because of the ban, Buehrle and his family live in Broward County, about thirty-eight miles away from Miami, and he had a longer commute to work. In 2013, Buehrle was traded to the Toronto Blue Jays and faced an even bigger problem since pit bulls are not legally permitted in any part of Ontario. Buehrle and his family made the difficult choice that he would live in Canada while his family and their dogs lived in St. Louis with his wife's family.

43 Yahoo! "The Pitbull Ban in Miami-Dade County Forces Mark Buehrle's Family to Settle Elsewhere" (January 6, 2012), https://sports.yahoo.com/blogs/big-league-stew/ miami-pitbull-ban-forces-mark-buehrle-family-settle-182451604.html.

Pit bull bans are so fervently enforced that they've even been applied to service dogs. In Denver, a Vietnam veteran suffering from post-traumatic stress disorder depended on his pit bull dog, Precious, to cope with his disability, but in 2012, city officials seized his dog, citing the city's pit bull ban. The city finally returned the dog to him, but he had to agree to keep the dog muzzled and comply with other restrictions. The courts found in favor of the veteran since the Americans with Disabilities Act prohibits flat-out breed discrimination of service dogs.

In 2011, a former Chicago police officer who had retired to Aurelia, Iowa, was suffering from disabilities resulting from a stroke and relied on his part pit bull, part Labrador service dog, Snickers, to help him. The city of Aurelia, however, told him that he could not keep the dog due to its pit bull ban. The city asserted that the Americans with Disabilities Act was not violated since the officer could simply get a new dog that did not violate the ban; a federal judge ruled against the city.

Montreal, Canada, passed a ban on pit bulls and "pit bull–type" dogs (whatever that means) in September 2016, and ordered the euthanasia of all such dogs by October 3, 2016. The ban, however, was immediately challenged in court by the Montreal SPCA, with support from the city's mayor, and was lifted in December 2017. In June 2018, Quebec did the same.

The United Kingdom's Dangerous Dogs Act 1991, banned pit bulls and three other breeds.

Pit bull bans are wide-ranging, and in many cases, the authorities need only believe that a dog appears to be a pit bull to have the right to seize it. In other words, if a dog looks like one, he is one and is therefore banned. Section 8.55 of the Denver Municipal Code, which was enacted in 1989 and is still applicable today, defines a pit bull as:

Any dog that is an American pit bull terrier, American
Staffordshire terrier, Staffordshire Bull terrier, or any dog
displaying the majority of physical traits of any one (1) or more
of the above breeds, or any dog exhibiting those distinguishing
characteristics which substantially conform to the standards
established by the American Kennel Club or United Kennel Club
for any of the above breeds. The A.K.C. and U.K.C. standards
for the above breeds are on file.[44]

The ability to ban an animal based on appearance and an
assumed corresponding personality is stereotyping and profiling
in the dog world. People who try to make their dog (of any breed)
vicious are the ones who should be prosecuted. Legislators,
enforcement personnel, and insurance companies should know
this and not succumb to discriminatory profiling. Lawmakers
and enforcement personnel are too quick to use a one-size-fits-all
mandate. According to the website DogsBite.org, nine hundred
US cities have enacted some sort of breed-specific legislation.[45]
The result is that people have to hide, get rid of, or euthanize their
beloved pets, and that countless hours have been spent in court
to both fight this and also defend dogs who are accused of being
pit bulls but are not. DNA testing can help prove a dog isn't a pit
bull, but this, unfortunately, tends to lose to the argument that
local authorities have jurisdiction over public safety issues and, if
they consider a particular dog unsafe, they can take it away.

Consider this scenario: a friend of mine adopted a dog from
a shelter. The adoption documents characterized the dog as a
boxer mix. My friend then took the dog to a veterinarian for a

---

44 *Colorado Dog Fanciers v. Denver*, 820 P.2d 644 (1991),
    https://law.justia.com/cases/colorado/supreme-court/1991/90sa342-0.html.
45 DogsBites.org, "Breed-Specific Laws State-By-State,
    https://www.dogsbite.org/legislating-dangerous-dogs-state-by-state.php.

checkup and vaccinations, and the veterinarian, unbeknownst to my friend, labeled the dog a pit bull mix in his paperwork. The dog was a wonderful dog and there were no problems until my friend found himself in a legal conflict with his landlord. When it was revealed that the dog was classified as a pit bull mix by the veterinarian, my friend was accused of having lied about the dog's breed in in the rental application to circumvent the "no pit bull" lease clause. My friend was evicted. This is what happens when dogs are judged by looks alone. DNA testing can help confirm breed in a situation like this, but the tests are not perfectly reliable. There is no objective way to know what a mixed breed of any kind, designer or not, has on the inside.

How does one explain to a child that his or her pet must be killed because the pet looks like a member of a breed that is considered vicious?

Threats of Extinction

While some breeds are banned, several others are at risk of extinction. Breeds such as the French bulldog, pug, boxer, Pekingese, shih tzu, and related designer crossbreeds, can no longer breed on their own. Their physical makeup has been so degraded by inbreeding, and they suffer from so many respiratory and stress issues, that it has become physically impossible for them to breed naturally. Artificial insemination followed by a caesarean section is the only way to go. Other breeds are simply being ignored into oblivion due to the sustained interest in designer dogs and teacups. In 2017, Britain's Kennel Club, the organizer of the Crufts dog show, expressed concern that small custom-designed dogs are threatening some famous British breeds with extinction as the registry of new puppies shows a dip in births of Skye terriers, bloodhounds, King Charles spaniels, and mastiffs. A March 7, 2017 article by Thea de Gallier titled "Crufts 2017: Will the world-famous dog show see FEWER British breeds this year?" and featured in England's *Express* reads:

> [There have been] warnings from the Kennel Club that three British dog breeds—the Skye terrier, the Otterhound and the Sussex Spaniel—are "as endangered as the Giant Panda."[46]

> The popularity of "handbag dogs" like Chihuahuas and breeds popular with celebrities, like pugs and French bulldogs, has been blamed for the decline in birth rates of traditional British breeds.

> Just 120 puppies from the three at-risk breeds were born last year, compared to 20,000 French bulldogs.

On the other side of the pond, the Westminster Kennel Club Dog Show introduced three new breeds in 2017, the American hairless terrier (billed as hypoallergenic), the sloughi, and the pumi, and, for the first time, exhibited cats as well.

What is interesting, though sad, is that this demonstrates the effectiveness of the theory of supply and demand. The lack of interest in some traditional breeds reduces the demand side, and the supply side responds in kind. Mishandling, shunning, banning, dismissing, and ignoring have resulted in breeds fading fast.

---

46 Thea de Gallier, "Crufts 2017: Will the World-Famous Dog Show See FEWER British Breeds This Year?" *Express* (March 7, 2017), https://www.express.co.uk/news/nature/776178/crufts-dog-world-famous-show-fewer-british-breeds.

More Valuable Than Gold

Since the designer dog obsession started about twenty years ago, we have seen an increase in incidents of stolen family pets. There are some estimates that nearly two million pets a year are stolen.[47] Thieves see people dressing up their dogs, traveling on vacation with them, dining out with them, and taking their dogs to work, and essentially everywhere else, and note how valuable the dogs are to their owners. People spend a fortune on dog toys and treats, day care, spa treatments (guilty), and pet sitters to pamper and ensure the comfort and safety of their canine family members. There are now even apps and gizmos that allow you to watch your dog and speak to him or her when you're away. Criminals have noted all this, and realized that designer dogs can be worth more than gold to some people. In other words, emotional importance = high monetary value.

As with everything trendy, stealing for resale is expected. And with designer dogs, it's no surprise that thieves use them to feed the black market. But the dognapping of designer dogs has led to a world of ransom demands, which can be more lucrative for criminals.

For some thieves, stealing a dog is easier and safer than committing other crimes. They can take a dog tied up outside a coffee shop, grab a pet crate that was put down for merely seconds, remove a dog from a backyard, steal a dog from a car, or grab a loosely held leash, much like a purse snatch. The United States and other countries, including the United Kingdom, Canada, and Mexico, are reporting meteoric rises in dognapping both for resale and ransom.

spcaLA regularly receives calls from distraught pet owners

---

47   Pet FBI, "Stolen Pets," https://petfbi.org.

who say their pet was stolen—grabbed from their hands, stolen from their car, grabbed away from their dog walker—and that they received a call from the perpetrator to pay a ransom for the dog's safe return. We were involved in a residential home burglary case where the only thing stolen was a litter of boxer puppies. The crimes in this case were grand larceny and extorting a ransom. Typically in these cases, we bring in additional law enforcement and advise the distraught owner not to meet the criminal without police backup. As with kidnapped children, frantic pet owners often disregard that directive for fear that the kidnappee will be hurt, and simply make the exchange. Others work with law enforcement, succeed in retrieving their pet, and facilitate the arrest of the dognapper. There are no guarantees that one approach will be more successful than another.

In Atlanta, in February 2017, thieves burglarized a man's home after seeing him walking his toy poodle. They stole the poodle and a variety of items. The man's roommate posted an ad for the stolen dog on Craigslist and got a call demanding a ransom. The dog is still missing.

In New Brunswick, Canada, in January 2017, a shepherd mix was stolen from a backyard. The alleged thief wanted the ransom wired to him and promised to return the dog after receiving it, but got spooked and hung up when he realized the police were listening to the phone call. It is not clear if he actually had the dog or was only claiming to, sharing details that he got from a missing-dog poster. The dog is still gone.

In the United Kingdom, in January 2016, two pregnant purebred dachshunds were stolen; a £3,000 ransom was paid for their return. In Oaxaca, Mexico, a Chihuahua was taken from someone's backyard. The owners could not afford the ransom, but offered to instead feed the dognapper whenever he wanted, since they owned a small restaurant. Eventually, however, a ransom was paid. In the United Kingdom, in February 2018, four labradoodle puppies, worth about £800 each, were stolen at knifepoint.

The number of dognappings that occur is unknown and is an underreported crime. I have read news accounts that allege that millions of dogs are stolen worldwide each year; other accounts estimate even more. Many of the crimes go unreported because people are afraid to involve law enforcement or are embarrassed to admit they paid a scammer. Additionally, when reported, the theft is listed as a property crime rather than a pet theft. What is clear is that it's important to be mindful and vigilant and not leave your pet unattended or speak in public about the price you paid for your pet. It's also important to be wary of anyone asking too many questions about your dog, and a good idea to carefully vet dog walkers and pet sitters.

Many criminals have figured out that when they see a "Lost dog: reward" sign, they can obtain the ransom without even taking the dog. For this reason, it's necessary to be careful when deciding what information to give. Too detailed a description can provide a scammer with enough information to fool you. Hold back on unique characteristics so you can verify if the person actually has your dog. Also be wary of con artists who might use social media photos of your dog to trick you into thinking they have it. Offering a reward also begs the question of how much is appropriate. Not enough and the thief might sell the dog for more to someone else. Too much and scammers line up to bid. One argument for promising a reward is that it might motivate the thief to demand a ransom and give the dog back quickly, rather than figure out where to try for a sale.

There is a difference of opinion about whether to provide a

dog's name on a sign, or even on a collar tag. The fear is that a friendly dog will respond to his or her name and walk toward the thief. On the other hand, if the dog is lost and a Good Samaritan sees the dog, knowing the name could help catch and secure him. Frankly, I don't put names on my dogs' tags or my children's backpacks for fear of them feeling safe if a stranger calls them by name.

Because many beloved lost and stolen pets end up for sale in assorted seedy places such as the backs of pickup trucks, flea markets, pubs, or street corners, it might give designer dog bargain-hunters pause to know that they are supporting these thieves, who will continue doing this as long as there is a demand. No good can come from buying a dog from the back of a truck.

Celebrities are not exempt (you already know the story of Paris Hilton's dog Tinkerbell who was stolen and ransomed for $5,000), and neither are pet stores. In February 2017, two designer dogs, a shih tzu-poodle mix and a Havanese, were stolen from a pet store in Arlington Heights, Illinois. The dogs are valued at more than $2,000 apiece. There are no reports that they've been found.

CHAPTER 12

THERE OUGHT TO BE A LAW

At his best, man is the noblest of all animals; separated from law and justice he is the worst.

　—ARISTOTLE

The enactment of laws can articulate goals and issue red cards to penalize those who cheat. Unfortunately, there aren't enough ethical referees to have an impact, and we don't set the rules for the entire world. In the area of closing puppy mills and punishing irresponsible breeders, there has been little improvement despite the laws passed, most of which were designed to protect animals from harm and reduce the demand for puppy mill dogs at the outset by restricting "people, places, and things" from participating in the process. The hope is that constructing one barrier at a time will ultimately have the cumulative effect of reducing the supply.

Who can sell or deal in the dog business and what can be sold are examples of the "people, places, and things" construct. Some laws offer the "how" when mandating humane care and treatment. The "why" is determined by ever-evolving shifts in policy, community standards, and cultural norms. Everything in this field is emotionally charged: from an impulse decision to buy a teacup dog after seeing a cute one on social media, to the frustration with the establishment's handling of the industry and the sorrow felt when a puppy suffers. Almost all evoke visceral responses.

The treatment of animals in laboratories is still suspect. The continued use of and need for dogs to be tested on by the biomedical community, the continued stealing of dogs by class B dealers to sell to laboratories, and the poor treatment of these dogs has not improved since 1966. The mass breeding of dogs has increased to accommodate the designer dog mania, and the value of the dogs has risen dramatically. This provides more incentives and opportunities for bad actors to steal and resell dogs.

There are two new federal bills proposed in Congress to address these issues. The first, H.R. 1141: Pet Safety and Protection Act of 2017, was introduced by Representative Michael Doyle on February 16, 2017. It is still in committee. It seeks to amend the Animal Welfare Act to "ensure that all dogs and cats used by research facilities are obtained legally."[48] The intention of the proposed plan is to stop class B dealers from selling to such facilities and to have an infraction penalty of $1,000 for breeders, pounds, and owners who violate the law. This significant bill would eliminate a class of dealer and institute a fine for misbehavior by others, which could be a meaningful deterrent, though there will always be people who break the law or sideline it.

As with any law, it is meaningless without committed enforcement. There has not been sufficient criminal enforcement of legislation promoting animal welfare since the Animal Welfare Act of 1966. Consumers who would like to help enforce it, by

---

48 115th Congress, "H.R.1141 - Pet Safety and Protection Act of 2017" (February 16, 2017), https://www.congress.gov/bill/115th-congress/house-bill/1141/text?format=txt.

shunning businesses that have complaints and violations in their treatment of animals, have been stymied by the decision that sanitized many of those sections on the USDA website.

I wonder if H.R. 1141 would make it more attractive for pounds to rethink their pound seizure positions since, in theory, fewer seller categories can result in more sales for the remaining sellers. This is also true of the class A breeders, who would have an incentive to increase their breeding operations.

The opening portion of H.R. 1141 reads:

To amend the Animal Welfare Act to ensure that all dogs and cats used by research facilities are obtained legally.

Be it enacted by the Senate and House of Representatives of the United States of America in Congress assembled,

SECTION 1. SHORT TITLE.

This Act may be cited as the "Pet Safety and Protection Act of 2017".

SEC. 2. PROTECTION OF PETS.

(a) RESEARCH FACILITIES.—Section 7 of the Animal Welfare Act (7 U.S.C. 2137) is amended to read as follows:

SEC. 7. SOURCES OF DOGS AND CATS FOR RESEARCH FACILITIES.

(a) DEFINITION OF PERSON.—In this section, the term 'person' means any individual, partnership, firm, joint stock company, corporation, association, trust, estate, pound, shelter, or other legal entity.

(b) USE OF DOGS AND CATS.—No research facility or Federal research facility may use a dog or cat for research or educational purposes if the dog or cat was obtained from a person other than a person described in subsection (d).

(c) SELLING, DONATING, OR OFFERING DOGS AND CATS.—No person, other than a person described in subsection (d), may sell, donate, or offer a dog or cat to any research facility or Federal research facility.

(d) PERMISSIBLE SOURCES.—A person from whom a research facility or a Federal research facility may obtain a dog or cat for research or educational purposes under subsection (b), and a person who may sell, donate, or offer a dog or cat to a research facility or a Federal research facility under subsection (c), shall be—

(1) a dealer licensed under section 3 that has bred and raised the dog or cat;
(2) a publicly owned and operated pound or shelter that—
   (A) is registered with the Secretary;
   (B) is in compliance with section 28(a)(1) and with the requirements for dealers in subsections (b) and (c) of section 28; [These refer back to the five-day holding periods, and dealers' compliance with certification requirements that the pets come from registered breeders and not random sources.] and

(C) obtained the dog or cat from its legal owner, other than a pound or shelter;

(3) a person that is donating the dog or cat and that—

    (A) bred and raised the dog or cat; or

    (B) owned the dog or cat for not less than 1 year immediately preceding the donation;

(4) a research facility licensed by the Secretary; and

(5) a Federal research facility licensed by the Secretary.

(e) PENALTIES.—

    (1) IN GENERAL.—A person that violates this section shall be fined $1,000 for each violation.

    (2) ADDITIONAL PENALTY.—A penalty under this subsection shall be in addition to any other applicable penalty.

(f) NO REQUIRED SALE OR DONATION.—Nothing in this section requires a pound or shelter to sell, donate, or offer a dog or cat to a research facility or Federal research facility.[49]

The second federal bill, H.R. 816: Federal Accountability in Chemical Testing (FACT) Act, introduced by Representative Ken Calvert on February 2, 2017, and also still in committee, seeks to improve reporting by assorted federal government agencies, such as the FDA and USDA, National Institutes of Health (NIH), and others, to report their efforts to utilize alternative test methods (i.e., not testing on animals) to assess the toxicology of cosmetics, drugs, foods, and other substances. Records documenting the numbers of animals, species, and tests performed would need to be reported so that Congress can determine if federal agencies are

---

49 Ibid.

truly striving to replace unnecessary animal testing with better alternatives. Note that "ICCVAM," mentioned below, stands for the "Interagency Coordinating Committee on the Validation of Alternative Methods." Better communication among agencies will increase efficiency and could have the added benefit of reducing some redundancies in animal testing. Currently, multiple laboratories perform the same tests and each uses animals, so multiple animals are used in the process. It would be more ethical if one laboratory performs the test and shares its results, so that fewer animals are used.

Here is the opening text of H.R. 816:

To amend the ICCVAM Authorization Act of 2000 to improve reporting about animal testing and alternative test method use by Federal agencies, and for other purposes.

Be it enacted by the Senate and House of Representatives of the United States of America in Congress assembled,

SECTION 1. SHORT TITLE.

This Act may be cited as the "Federal Accountability in Chemical Testing Act" or the "FACT Act".

SEC. 2. REPORTING REQUIREMENTS; PROVISION OF INFORMATION.

(a) REPORT CONTENTS.—
(1) IN GENERAL.—Paragraph (7) of section 3(e) of the ICCVAM Authorization Act of 2000 (42 U.S.C. 285l–3(e)) is amended by inserting before the period at the end

of the paragraph the following: "to include a description of the progress on the development, validation, acceptance, and utilization of alternative test methods (including animal use data by species, number, and test type) for toxicological testing conducted, supported, or required by, or submitted to, each Federal agency listed in subsection (c) during the reporting period."

(2) EFFECTIVE DATE.—The amendment made by paragraph (1) applies with respect to reports that are made available to the public on or after the date that is 90 days after the date of enactment of this Act.

(b) PROVISION OF INFORMATION.—Section 4 of the ICCVAM Authorization Act of 2000 (42 U.S.C. 285l–4) is amended by adding at the end the following:

(f) PROVISION OF INFORMATION.—Each Federal agency carrying out a program described in subsection (a), or its specific regulatory unit or units, shall provide ICCVAM with information needed to carry out this Act.[50]

This bill mandates that reports of agencies testing on animals be available to the public, on the USDA's website, which will circumvent the existing whitewash of the USDA site.

---

50 115th Congress, "H.R.816 - FACT Act" (February 2, 2017), https://www.congress.gov/bill/115th-congress/house-bill/816/text?q=%7B%22search %22%3A%5B%22actionDateChamber%3A%5C%22115%7CH%7C2017- 02-02%5C%22+AND+%28billIsReserved%3A%5C%22N%5C%22+or+type %3A%5C%22AMENDMENT%5C%22%29%22%5D%7D&r=11.

Combating False Advertising

In keeping with the spirit of the federal requirement that the prospective pet and purchaser be in the same place at the same time, to avoid sight-unseen surprises using bait-and-switch advertising techniques, states have been aggressive in "enforcing similar existing local laws, or enacting new ones." California, for example, introduced a bill in 2017, A.B.: Sale of dogs or cats, that would criminalize the false advertising, in any medium, of dogs and cats. This is necessary because no one intentionally chooses a sick, near-death dog to purchase. But that is often what people receive when they rely on advertising photos and statements that make the dog seem healthy and sane. If breeders can no longer sell sick dogs, maybe they would put resources into keeping the ones they have healthy.

Proposed language from the bill reads:

SECTION 1.

Section 17531.2 is added to the Business and Professions Code, to read:

17531.2.

(a) It is unlawful for any person, firm, or corporation, in any newspaper, magazine, circular, form letter, or any open publication, published, distributed, or circulated in the State of California, including over the Internet, or on any billboard, card, label, or other advertising medium, or by means of any other advertising device, to advertise, call attention to, or give publicity to, the sale or transfer of a dog or cat for which any of the following apply:

(1) The advertised dog or cat is not actually available for purchase by the public at the time of the advertisement or offer.

(2) Pictures or descriptions of a dog or cat are not of the dog or cat actually available for purchase by the public at the time of the advertisement or offer.

(3) Statements about the dog or cat being advertised or offered for sale are known, or by exercise of reasonable care should be known, to be untrue or misleading.

(4) Statements about the dog or cat are made without the actual intent to sell or offer the exact dog or cat so advertised or offered.

(b) In addition to any other penalty provided by law, any violation of this section is a misdemeanor punishable by imprisonment in the county jail not exceeding six months, or by a fine not exceeding two thousand five hundred dollars ($2,500), or by both that imprisonment and fine.[51]

Laws like this protect consumers and pets when the transaction is intrastate, and a federal statute will not. Even though all states already have false advertising and larceny laws on the books, a law like this emphasizes that this applies to pets too. Unfortunately, at the end of 2017, the proposed law was vetoed by the governor of California due to this redundancy.

Retail pet sourcing laws are aimed at controlling the source of animals from which a pet shop can build its inventory. There is a national trend to prohibit the sale in pet shops of

---

51 State of California, Business and Professions Code, Section 1.Section 17531.2, https://leginfo.legislature.ca.gov/faces/billCompareClient.xhtml? bill_id=201720180AB1138.

commercially bred, puppy mill dogs or dogs from other breeders. The thinking here is that if selling puppy mill dogs in retail outlets is prohibited, there will be fewer places to sell designer dogs. This would mean that fewer people would be exposed to them, so sales would decline and suppliers would cut back production in response. These bans would not stop individuals from purchasing directly from puppy mills, but it would add the extra work of having to do so and cut out people who don't want to put in the effort.

There are isolated communities, such as Manhattan Beach, California, that have local ordinances that ban the sale of live animals in retail establishments. Stores in these communities sell pet foods, treats, and accessories, and often host adoption events, during which animal shelters showcase, and hopefully place, dogs into homes. These events help shelters and their animals, help the host stores sell supplies for new pets, and help the public feel secure in knowing that they are helping pets in need.

Today, many pet shops that sell dogs only sell shelter and rescue dogs. In 2000, spcaLA, at the forefront of this trend, opened its first mall store and filled it with pet supplies and shelter pets. Unlike bringing pets to a store for a weekend or having animals in a store for a Christmas holiday event, the spcaLA store was a real year-round store in a mall. We were fortunate enough to be given the space at no charge, except for some incidental mall-related fees. Our theory was that the products sold in-store would support the cost of the personnel needed to run the store and conduct adoptions. Of course, the store had to abide by mall rules, keep mall hours, and operate at the highest level of attractiveness, so we had to focus on aesthetic appeal, customer service, and controlling odors and debris. (My first job at our mall store was to clean the windows.)

In doing so, we successfully created the first pet shop that carried only shelter animals. In fact, the store was so popular and so welcomed by the community and mall shoppers that we soon had five such stores in five different malls around town. In some locations, the spaces were large enough to offer training classes and a variety of lectures and fundraising events. The theory that we could successfully place shelter dogs through this method was working and we were breaking even, so the model seemed successful and replicable. One store at a mall in Manhattan Beach was considered a community shelter and thus exempted our animals from their animals ban.

By bringing the animals into malls, we reached people who may not have gone to a shelter. Malls also didn't need to lease the spaces to overcrowded pet stores that were only interested in profits and viewed animals as non-sentient inventory. spcaLA stores demonstrated the value of treating animals humanely. This was crucial, since millions of children frequented the malls and we were determined to convey the importance of being kind to and mindful of animals.

After a decade each of the malls closed for renovation and when reopened could no longer afford to donate such generous square footage. New zoning and other laws have since been enacted as well, which makes the concept more difficult to execute. New rules against leaving the animals overnight would necessitate a host of logistical acrobatics to keep a store running. It has also become highly unusual for malls to donate space. We still partner with them for special adoption events and fundraisers, but we have moved on to new programs.

PetSmart has a program in which they build mini–brick-and-mortar animal shelters in their stores and partner with reputable shelters to stock and staff them. They call these Everyday

Adoption Centers, and they exist all over the country; spcaLA is the partner at one in California.

The concept of retail shops partnering with charities to provide animals in stores has gained traction and is being implemented in the form of retail pet ban laws. These laws essentially mandate that only animals from public and nonprofit shelter organizations be available for sale in pet shops. The stores can still be commercial for-profit establishments, but they're prohibited from selling dogs, cats, and rabbits that came from puppy and kitten mills and other breeders. The exact laws vary from city to city and state to state. Some states, such as New York, also allow dogs from class A breeders to be sold in pet stores.

For us animal welfare activists, the hope is that the demand for newly bred designer pets will diminish, causing animals already for looking for homes to find them and in the process ending the need to euthanize otherwise adoptable animals. This is noteworthy: because many designer dogs end up in shelters, there is a chance that the customer will leave with the specific breed of dog they wanted from the get-go. (There will be no purebred registration papers and the dog will not be able to breed due to spay and neuter mandates, but they can still have the dog they want.)

Currently, 196 cities in sixteen states have enacted some form of a retail pet sourcing law. In 2017, California enacted the first state law mandating that the only animals that can be sold in pet stores are dogs, cats, and rabbits from shelters, humane societies, and nonprofit animal rescue groups. This law will go into effect January 1, 2019. In April 2018, Maryland became the second state to pass a similar law. As of this writing, Pennsylvania, New York, and New Jersey are considering doing the same. Imagine the impact if there was a national law of this kind, particularly if it criminalized violations. Shelters would have more outlets to place animals, which would free up cage space and allow more time to find homes for their residents. Potential pet parents would have more convenient locations to seek a new family member. Pet shops would have no reason to order pets from puppy mills and other unscrupulous breeders. And because of all of this, pet overpopulation and the euthanasia that results could become a thing of the past! How wonderful would that be?

There are important steps to be taken in helping to inform the nation of the sad fate of unwanted animals and modify the thinking that encourages a revolving-door attitude toward pets, and the waste of so many lives. That said, these are not simple steps; there are many details that need to be thought through. It is important that we ensure that the enactment of these ordinances be a thoughtful, enforceable, tangible step toward a goal, rather than a political chimera. The creation of shelter contracts with for-profit stores, involving fees, length of stay, need for staff, transfers, and enforcement of humane protections can be complicated. Additionally, it is critical to thoroughly vet nonprofit groups if they are not established organizations. We know some are illegitimate and, as previously mentioned, some breeders are forming their own nonprofits and using them to place their dogs in stores, ostensibly to comply with the law. As of this writing, there are multiple criminal investigations into such entities for this kind of violation. A nonprofit with an unlimited supply of designer puppies to provide to stores should sound an alarm. One such group is falsely claiming that its supply consists of the runts of the litters and those puppies not eligible for show, due to an imperfect appearance. There is also an effort to bypass stores and reach out directly to consumers to purchase these dogs.

Lawsuits around the country have been brought on behalf of pet shops that claim that laws that limit where their supply comes from hinder the free flow of interstate commerce, but so far they have been unsuccessful. So too have lawsuits that claim that states that require animals to be sterilized before leaving the shop burden the flow of commerce. Sterilization requirements are important as they prevent people from buying animals in pet stores for breeding purposes. The laws across the country are not uniform on this.

In the lawsuit *New York Pet Welfare Association, Inc.* v. *New York City*, which was decided on March 2, 2017, the New York Pet Welfare Association, which represented pet shops and breeders, claimed that the state law governing the sale and sterilization of pets is preempted by federal law, and that mandatory sterilization of sold pets and restrictions on the source of pets placed burdens on interstate and state commerce. The case was brought to prevent a New York law written in 2015 from taking effect. That law required New York pet shops to only purchase from federally licensed Class A dealers, not Class B random source dealers, and also to sterilize each animal before releasing it to a buyer.

The United States Court of Appeals for the Second Circuit, a federal court, ruled against the pet shops and breeders, and addressed the main issues beautifully, clearly, and rationally. First, regarding the issue of state law being preempted by federal law, in this case the Animal Welfare Act, and the association's claim that the former could not mandate any different than the latter, the United States Court of Appeals, a federal court, held that this was not so. In its ruling, the court wrote:

> Significantly, the text of the AWA unambiguously envisions continuing state animal welfare regulation. Congress has

expressly authorized the Secretary, in implementing the Act, to "cooperate with" local officials "in carrying out the purposes of ... [the AWA] and of any State, local, or municipal legislation or ordinance on the same subject." ... The Secretary may preempt only to the extent that Congress has delegated him the power to do so. ... [The AWA's] Section 2145(b)'s explicit grants of authority to the Secretary to cooperate with state officials carrying out state law makes it clear that Congress did not intend that the statute displace all state regulation of the field.[52]

In other words, the court felt that it was always the intent of the Animal Welfare Act to let states and local municipalities regulate animals as they see fit. The judge specifically pointed out that the Secretary of Agriculture is authorized to cooperate with the states in effectuating the purposes of this act.

Then there's the question as to whether the retail source ban discriminates against interstate commerce, thus impeding it, or whether it is a nondiscriminatory regulation with only incidental effects on interstate commerce. In this context, discrimination meant treating in-state and out-of-state players differently, so that there was a benefit for the in-state player and a burden on the out-of-state player. The discrimination was alleged to have occurred because out-of-state dealers couldn't use local class B middleman, so the effect of the law shifted the purchase of animals from dealers outside New York to animal shelters inside New York. The court ruled no problem here:

---

52 United States Court of Appeals, Second Circuit. *New York Pet Welfare Association Inc.* v. Linda Rosenthal Inc. (March 2, 2017), https://caselaw.findlaw.com/us-2nd-circuit/1850978.html.

The City has identified a number of local benefits that are clearly unrelated to economic protectionism . . . Requiring pet shops to purchase directly from Class A breeders protects consumers by making it impossible to obscure the source of an animal by using a middleman, enhances animal welfare by reducing the incidence of disease and behavioral problems associated with irresponsible breeding, and alleviates the burden of providing care in public shelters for animals abandoned because of such problems.[53]

The court found that the law did not discriminate against out-of-state players. Regarding the final issue, whether complying with a mandatory sterilization before release is preempted by other state law, the court said the claim was nonsense. The implemented New York law sets the minimum age to sterilize a dog or cat at eight weeks and the minimum weight for this as two pounds, and requires that it be done while the animal is still in the custody of the pet store. In response to the pet stores' argument that required sterilization violates a veterinarian's duties under New York law to (1) apply independent professional judgment, and (2) obtain the informed consent of the owner before performing a surgical procedure, the judge said that if a veterinarian felt uncomfortable doing the surgery they didn't have to, but the store needed to find another who would.

Laws have also been enacted around the country that address where pets may be sold besides a pet store. Effective in 2016, a part of the California Penal Code 597.4 that prohibits the sale of animals on the street reads:

---

53 Ibid.

(a) It shall be unlawful for any person to willfully do either of the following:

   (2) Sell or give away as part of a commercial transaction a live animal on any street, highway, public right-of-way, parking lot, carnival, or boardwalk.

   (3) Display or offer for sale, or display or offer to give away as part of a commercial transaction, a live animal, if the act of selling or giving away the live animal is to occur on any street, highway, public right-of-way, parking lot, carnival, or boardwalk.

(b) (1) A person who violates this section for the first time shall be guilty of an infraction punishable by a fine not to exceed two hundred fifty dollars ($250).

   (2) A person who violates this section for the first time and by that violation either causes or permits any animal to suffer or be injured, or causes or permits any animal to be placed in a situation in which its life or health may be endangered, shall be guilty of a misdemeanor.

   (3) A person who violates this section for a second or subsequent time shall be guilty of a misdemeanor.[54]

Swap meets and other public places that sell assorted merchandise objected to this prohibition as they too were selling animals. We intentionally meant to include them in this prohibition, since animals sold at swap meets frequently are not handled humanely and are often not properly sheltered from the elements. After much negotiation with the swap meet groups and other licensed street sellers, California Health and Safety Code 122370

---

54 State of California, "California Code, Penal Code - PEN § 597.4," https://law.justia.com/codes/california/2016/code-pen/part-1/title-14/section-597.4/.

was enacted. It permits swap meet vendors to sell animals in public as long as the local jurisdiction has adopted standards of care for the species being sold. Text from the code reads:

> A swap meet operator may permit a vendor to offer animals for sale at a swap meet provided the local jurisdiction has adopted standards for the care and treatment of those animals during the time that the animals are present at the swap meet and transported to and from the swap meet. This chapter does not apply to the sale of a particular species of animal if a local jurisdiction has adopted a local ordinance prior to January 1, 2013, that applies specifically to the sale of that particular species of animal at swap meets.

> *(Added by Stats. 2013, Ch. 231, Sec. 1. Effective January 1, 2014. Section operative January 1, 2016, pursuant to Section 122374.)*

> **122372.** Any ordinance adopted pursuant to Section 122370 shall, at a minimum, require the swap meet vendor to do all of the following:

> (a) Maintain the facilities used for the keeping of animals in a sanitary condition.
> (b) Provide proper heating and ventilation for the facilities used for the keeping of animals.
> (c) Provide adequate nutrition for, and humane care and treatment of, all animals that are under his or her care and control.
> (d) Take reasonable care to release for sale, trade, or adoption only those animals that are free of disease or injuries.

(e) Provide adequate space appropriate to the size, weight, and species of animals.

(f) Have a documented program of routine care, preventative care, emergency care, disease control and prevention, and veterinary treatment and euthanasia that is established and maintained by the vendor in consultation with a licensed veterinarian employed by the vendor or a California-licensed veterinarian, to ensure adherence to the program with respect to each animal. The program shall also include a documented onsite visit to the swap meet premises by a California-licensed veterinarian at least once a year.

(g) Provide buyers of an animal with general written recommendations for the generally accepted care of the type of animal sold, including recommendations as to the housing, equipment, cleaning, environment, and feeding of the animal. This written information shall be in a form determined by the vendor and may include references to Internet Web sites, books, pamphlets, videos, and compact discs.

(h) Present for inspection and display a current business license issued by the local jurisdiction where the animals are principally housed.

(i) Maintain records for identification purposes of the person from whom the animals offered for sale were acquired, including that person's name, address, e-mail address, and telephone number and the date the animals were acquired.

(Added by Stats. 2013, Ch. 231, Sec. 1. Effective January 1, 2014. Section operative January 1, 2016, pursuant to Section 122374.)

122372. (a) (1) A swap meet vendor who offers animals for sale at a swap meet in a local jurisdiction that has not adopted an ordinance authorizing that sale, is guilty of an infraction punishable by a fine not to exceed one hundred dollars ($100).

(2) A swap meet vendor who violates paragraph (1) for a second or subsequent time, is guilty of an infraction punishable by a fine not to exceed five hundred dollars ($500) per violation. The court shall weigh the gravity of the violation in setting the amount of the fine.

(3) Nothing in paragraph (2) shall preclude punishment under any other provision of law, including, but not limited to, laws prohibiting the abuse or neglect of animals in the Health and Safety Code or the Penal Code.

(b) A notice describing the charge and the penalty for a violation of this section may be issued by any peace officer, animal control officer, as defined in Section 830.9, or humane officer qualified pursuant to Section 14502 or 14503 of the Corporations Code.

(Added by Stats. 2013, Ch. 231, Sec. 1. Effective January 1, 2014. Section operative January 1, 2016, pursuant to Section 122374.[55]

The following two California Penal Code statutes are examples of laws that govern the euthanasia of a newborn animal, and that prohibit the sale of underage dogs. We have used this section on more than one occasion to convict a person who killed a newborn inhumanely. Puppy mill operators and pet shop owners who choose not to provide veterinary care to a sick animal will

---

55 State of California, "California Code, Health and Safety Code - HSC § 122370," https://codes.findlaw.com/ca/health-and-safety-code/hsc-sect-122370.html.

certainly not pay to put one to sleep humanely. To them it doesn't make financial sense.

No person, peace officer, officer of a humane society, or officer of a pound or animal regulation department of a public agency shall kill any newborn dog or cat whose eyes have not yet opened by any other method than by the use of chloroform vapor or by inoculation of barbiturates. . . .

(1) Except as otherwise authorized under any other provision of law, it shall be a crime, punishable as specified in subdivision (b), for any person to sell one or more dogs under eight weeks of age, unless, prior to any physical transfer of the dog or dogs from the seller to the purchaser, the dog or dogs are approved for sale, as evidenced by written documentation from a veterinarian licensed to practice in California.

(2) For the purposes of this section, the sale of a dog or dogs shall not be considered complete, and thereby subject to the requirements and penalties of this section, unless and until the seller physically transfers the dog or dogs to the purchaser.

(b)(1) Any person who violates this section shall be guilty of an infraction or a misdemeanor.

(3) An infraction under this section shall be punishable by a fine not to exceed two hundred fifty dollars ($250).

(4) With respect to the sale of two or more dogs in violation of this section, each dog unlawfully sold shall represent a separate offense under this section.

(c) This section shall not apply to any of the following:

   (1) An organization, as defined in Section 501(c)(3) of the Internal Revenue Code, or any other organization that provides, or contracts to provide, services as a public animal sheltering agency.[56]

Finally, I have mentioned cruelty in animal statutes several times in this book. The following section of the California Penal Code is its basic foundational prohibition against harming animals. Every state has one that is a very close version of this. It reads:

(a) Except as provided in subdivision (c) of this section or Section 599c, every person who maliciously and intentionally maims, mutilates, tortures, or wounds a living animal, or maliciously and intentionally kills an animal, is guilty of a crime punishable pursuant to subdivision (d).

(b) Except as otherwise provided in subdivision (a) or (c), every person who overdrives, overloads, drives when overloaded, overworks, tortures, torments, deprives of necessary sustenance, drink, or shelter, cruelly beats, mutilates, or cruelly kills any animal, or causes or procures any animal to be so overdriven, overloaded, driven when overloaded, overworked, tortured, tormented, deprived of necessary sustenance, drink, shelter, or to be cruelly beaten, mutilated, or cruelly killed; and whoever, having the charge or custody of any animal, either as owner or otherwise, subjects any animal to needless suffering, or inflicts unnecessary cruelty

---

56 State of California, "California Code, Penal Code - PEN § 597v," https://law.justia.com/codes/california/2015/code-pen/part-1/title-14/section-597v/.

upon the animal, or in any manner abuses any animal, or fails to provide the animal with proper food, drink, or shelter or protection from the weather, or who drives, rides, or otherwise uses the animal when unfit for labor, is, for each offense, guilty of a crime punishable pursuant to subdivision (d) [the penalty section].

(a) Every person who maliciously and intentionally maims, mutilates, or tortures any mammal, bird, reptile, amphibian, or fish, as described in subdivision (e), is guilty of a crime punishable pursuant to subdivision (d).

A violation of subdivision (a), (b), or (c) is punishable as a felony by imprisonment pursuant to subdivision (h) of Section 1170, or by a fine of not more than twenty thousand dollars ($20,000), or by both that fine and imprisonment, or alternatively, as a misdemeanor by imprisonment in a county jail for not more than one year, or by a fine of not more than twenty thousand dollars ($20,000), or by both that fine and imprisonment.[57]

---

57 Ibid.

CHAPTER 13

GIVE GENES A CHANCE

The ultimate test of man's conscience may be his willingness to sacrifice something today for future generations whose words of thanks will not be heard.

—GAYLORD NELSON

Laws help, but they are not enough. Shelters have individually and collaboratively launched awareness and educational campaigns to highlight the painfully high cost of cute and create programs aimed at reducing the demand for puppy mill dogs by focusing attention on existing dogs. Coupled with law enforcement, they make a difference.

Another way to assist law enforcement is to educate and enlist community members to look out for certain problems. spcaLA, in conjunction with the State Humane Association of California (now called California Animal Welfare Association), California Animal Control Directors Association, and Born Free USA compiled a pet shop inspection guide designed to inform pet shop operators of their legal obligations in husbandry and care, and to provide basic instructions for anyone to inspect a pet shop. Consumers can do the inspections, document their findings, and report them to authorities. This works well, as customers appear to be browsing and we have spies where we need them. But we would never ask a civilian to confront store personnel. I was once at a pet shop raid with a news crew, when the owner of the store ran out and punched the cameraperson. Confrontation by a member of the public is not an option.

Air Chihuahua

The Air Chihuahua program was designed to redistribute the existing supply of dogs based on demand, on a national level, and discourage turning to a breeder to obtain a specific dog, thereby reducing the need to breed.

By melding the laws of supply and demand with the practices of nonprofits, rapid change on a grand scale can be achieved.

This country is not consistent in its dog demographics; there is no standard mix of dogs, large or small, designer or mutt, in every location. The general shelter population reflects trends by housing the trendy discards and surpluses, as in the case of dalmatians, but by and large cities don't share the same dog demographics. When I lived in New York City, the shelters I visited had mostly large shepherd-mix type dogs (dogs deemed too large for small apartments). A small purse dog was an anomaly. When displaced Hurricane Katrina dogs were airlifted to Los Angeles, the airplane was mostly filled with beagles and basset hounds (dogs more commonly used in hunting). I had no idea that those dogs sing constantly throughout the night, as they did while we were processing them. Fortunately, the equally amazed neighborhood sang along. Here in Los Angeles, we are overrun with Chihuahuas and purse dogs. (We can thank our celebrity-obsessed culture for that.)

These differences are interesting, but they also limit choices and can prevent clients from finding the dog they are looking for. Cages filled with tan Chihuahuas don't provide a diverse range of options for California dwellers, but someone in Denver who doesn't have easy access to one might cherish finding one. Limited options at shelters cause people to go to pet shops or order dogs over the internet. That doesn't help reduce the demand for new dogs, nor does it help shelter dogs find a home.

On New Year's Eve 2009, spcaLA launched its Air Chihuahua program, sending a planeload of Chihuahuas to the Dumb Friends League shelter in Denver, Colorado. In Los Angeles we were struggling with a glut of tan Chihuahuas and other little dogs, while my colleagues in Denver were lamenting that they had only large dogs. We shared the concern that our customers would leave shelters without a dog, and worse, would go to a breeder. (Customers couldn't just go to another nearby shelter to find their dream dog, since the problem was regional.) It was then that the idea of Air Chihuahua was born. The cost of flights, crates, toys, and snacks were donated to spcaLA, and the first planeload of Chihuahuas left for Denver. George Lopez of *Beverly Hills Chihuahua* fame heard about the New Year's Eve airlift and covered the event on his then-airing television talk show. The coverage was hilarious, mocking a celebrity red carpet walk to the plane, but his message was serious. Keeping it light, he blamed himself for causing a mad rush for the dogs and asked his audience to adopt instead, as he had. He has a couple of Chihuahuas.

The news media in both California and Colorado helped tell the story, so adopters in Denver knew the dogs were arriving and were waiting at the shelter there to adopt. The dogs who were not being adopted in Los Angeles instantly received homes in Denver. In our care, they could have waited months for a home; in Denver, they could be adopted in days. We got dogs to people in Denver, who otherwise would have caused puppy mill and backyard breeders to create more of them, and saved the new dog owners the exorbitant cost they were going to pay. Altogether, we joined forces and flew thirty-five Chihuahuas from Los Angeles to Denver. We continue to fly little dogs all over the country and sometimes to Canada, and the success has been repeated each time. Our program flies the dogs, rather than subject them to long road trips.

Imagine if this was done routinely by every shelter in the country. The supply of *existing* homeless pets would be relocated to where the demand exists, thus eradicating the market for those who would abuse animals for profit. We would find homes for existing pets while simultaneously drying up the need for unprincipled breeders to produce more. Intelligently managing this would rapidly reduce the shelter populations and the euthanasia rates, and it would satisfy clients. This concept was based on the for-profit sector; for example, a chain of clothing stores moved existing clothes from a region where they didn't sell to one where they were selling out, before ordering the manufacture of new items. With clothes, people can enjoy new styles. With animals, the difference can mean life and death.

An increasing number of leaders of legitimate shelters across the country are participating, which makes it easier for us to pool resources to fund it. Together it might one day be possible to find every adoptable pet a good home—wherever that may be. That was our New Year's resolution in 2009 and it continues to be our resolution each year.

There have been less successful copycat programs, less successful because they have a "just get them out of here" core attitude and the lack of passion produces paltry results. This has sparked criticism and an assertion that transporting animals from one place to another simply reassigns the task of euthanasia to the receiving entity and spreads disease to new populations. I can't speak to the internal workings of all the transport programs that have materialized since we began Air Chihuahua, but the accusation could have merit in some cases.

That is not the case with our program or any reliable program that I am familiar with. Our program has been incredibly well thought out. Here are some reasons for its success:

- The receiving entity must be legitimate and actually want the animals. We send small dogs to locations that don't have any but have a demand for them. The result is that there is usually a line of people at the shelter waiting to adopt them before the plane reaches the gate, and they find homes very quickly.

- The arrival of new dogs has not been found to accelerate the euthanasia of existing dogs at those locations. The new arrivals are placed too quickly to cause space challenges, and/or the organization has planned the request for the dogs to coincide with the availability of extra space.

- Supplying a shelter with a lot of small dogs will not take interest away from the larger ones. People partial to larger dogs will continue to visit them at the shelter and rarely convert to small dog people.

- Shoppers who are visiting the shelter because they heard it has small dogs may never have been to a shelter before. Some will become new supporters of animal rescue and see the important work that shelters and our program are doing, and spread word of this.

- We send very adoptable dogs who are simply not in demand due to our demographics, which affords our remaining dogs more time and resources to be rehabilitated and placed.

- The people who adopt small-breed dogs from the airlifts did not have to go to a breeder, a puppy mill, or an internet retailer to get a dog because their desired breed could not be found at a local shelter. What could be better than that?

The caveat, as always, is to work with credible and ethical partners so that the end result is not relocation, buck-passing, or statistical judo, but rather a reduction in overall pet overpopulation and euthanasia, and an increase in permanent adoptions.

Teach Your Children Well

Education is critical to stopping the abuse of animals. It is essential that shelters flood the airwaves and social media with two types of information: first on the cruelty of puppy mills and puppy trafficking, and the second on the advantages and rewarding experience of adopting a shelter dog. Educating the public and our children as to *why* that is, is the secret ingredient to a successful and humane future. The information must be heard, believed, and then owned by the listener, otherwise it is like a hollow command. If people understand what is happening and don't like it, they make different decisions.

There are suites of laws—national and local, existing and forthcoming—that are intended to control the flow of sick puppies around the world. The laws are intended to prohibit puppy mill dogs from being sold in pet stores, forbid the sales of dogs on sidewalks, at swap meets, and generally in public places, and make it a crime to sell underage dogs. There are also laws pending that will prevent certain financing contracts and false advertising practices, which if passed will cripple the class B random source dealers. But laws are not enough. They get circumvented, become dormant, and often are not enforced. The laws usually work on already law-abiding people, but not on the lawbreakers or the ignorant.

Laws also are not the most efficient messengers to effectuate lasting change in behavior. If we arrest one person for animal cruelty, he might go to jail, but will he re-offend or ultimately change his ways? I submit to you that he will re-offend and not change his ways. Educating and messaging reaches more people more often and is more effective. If those who hear the message understand it and explain it to someone else, we can effect lasting change. *Lasting* change, because making an informed decision that is meaningful will endure, whereas doing something under duress or threat of getting caught and punished will last only as long as the threat.

Transport programs like Air Chihuahua, adoption festivals with interactive booths, fun bonding events like dog walks, dog fitness classes, and training classes to improve behavior and communication with your dog, as well as agility and flyball activities, all serve to highlight the shelter dog and strengthen the human-animal bond with any dog. Once that bond is felt, it becomes more difficult to support an industry that hurts dogs.

Patronizing and encouraging shady breeders is like voting against our own interests, in which we love dogs and want them, therefore we allow them to be hurt so we can have them.

As for what we pass to the next generation, mentoring better behavior teaches more than lecturing. Our kids are watching us and learning the good, the bad, and the stupid. By exposing children to dog trading days, dogfights, and a puppy mill full of the sick, the suffering, and the dying, or to domestic violence, cheating, and stealing, we cause them to become desensitized and emulate the bad behavior The converse is of course true too; they will also emulate the good.

Just Watch the Movie

It is hard to resist the appeal of a dog actor who is adorable and appears able to do anything a human can do. It stands to reason that after watching *101 Dalmatians*, many people would want a dalmatian. Imagine my reaction, though, when someone turned a dalmatian into the shelter complaining that the dog did not behave the way Pongo did in the movie. Dogs don't make coffee and use a word processor in real life!

In real life, thanks to the movie, the now-overbred dalmatian is often deaf, high-strung, and difficult to control. They are assumed to be hard to train and incorrigible, since their human companions frequently don't realize that they're not responding to verbal instructions because they are deaf. A dalmatian will never type your term paper, but there are training techniques that work very well with deaf dogs or dogs who only understand a different language, like some shipped from overseas. It is a method that uses visual and environmental cues to signal a behavior request. But people often don't want to bother with this extra effort, so shelters end up with the dogs they consider to be stupid.

Of course, snuffing the demand to mass-produce the breed in the first place can prevent deafness at the start. Watch the movie, but don't buy the dog. After each film release, opportunistic breeders furiously begin manufacturing dalmatians. Looking out my office window after the film's release, I often saw vehicles driving by with cardboard signs that read "Dalmatians 4 sale—Cheap." The fact is that our shelter was already full of dalmatians shortly after the film's release. Many had health and behavioral issues and not one could use a coffee machine as Pongo did in the movie. I remember walking through our dog cottages and seeing polka dots everywhere. The same *101 Dalmatians* syndrome is being repeated with huskies due to the popularity of *Game of Thrones* and the resemblance of huskies to the dire wolves featured in that series. Huskies are being purchased and dumped at such an alarming rate that, in August 2017, one of the show's stars, Peter Dinklage, urged people to stop, saying:

Please, to all of *Game of Thrones*'s many wonderful fans, we understand that due to the direwolves' huge popularity, many folks are going out and buying huskies. Not only does this hurt all the deserving homeless dogs waiting for a chance at a good home in shelters, but shelters are also reporting that many of these huskies are being abandoned—as often happens when dogs are bought on impulse, without understanding their needs. Please, please, if you're going to bring a dog into your family, make sure that you're prepared for such a tremendous responsibility and remember to always, ALWAYS, adopt from a shelter.[58]

Instead of creating this demand, why not explain to your children that the dog in the film is a trained actor who has a home and that there are plenty of dogs in the shelter that need homes? Also, that you don't have to get the same dog that everybody else

58 People for the Ethical Treatment of Animals, "Peter Dinklage Asks 'Game of Thrones' Fans to Stop Buying Huskies" (August 15, 2017), https://www.peta.org/media/news-releases/peter-dinklage-asks-game-thrones-fans-stop-buying-huskies/.

has? Note though that it does no good to explain to children why they can't have the dog in the film when grown-ups are racing to order Chihuahuas, French bulldogs, multipoos, and teacup dogs because a celebrity has one. It is the same thing, and the kids will see right through it. Celebrities go clubbing with purse dogs, drive with dogs on their laps, dye their dogs' hair to match their own outfits, and engage in other antics that we don't have to emulate. (In 2008, the California legislature supported a bill dubbed the "Paris Hilton Law" that would have prohibited the very dangerous practice of driving with a dog, or anything, on your lap. It didn't pass.)

In *Legally Blonde*, a Chihuahua went to law school; in *Beverly Hills Chihuahua*, a dog was dognapped and sent to participate in dogfights in Mexico. These were not real events, the films were fictional entertainment, yet they resulted in a massive glut of Chihuahuas. Some are suffering the effects of poor breeding, some are very ill and teacup-sized, and many are homeless. The West Coast shelters have been saturated with tan Chihuahuas, directly attributable to the Hollywood effect.

It happened in a nanosecond and will take decades to fix, especially since puppy mills are making more as we struggle with an existing glut. For the fix to occur, future generations must come on board. Learning to differentiate between real and make-believe and to appreciate entertainment without copying every little thing requires critical thinking skills and positive mentoring. With the ability to click on a video and order the same jewelry or dress that a celebrity owns, it becomes the responsibility of the adult in the room to differentiate a necklace from a dog.

Making the effort to explain and demonstrate moral behavior will go far in shaping our kids. In fact, it is not uncommon for our children to come home from school, camp, or a friend's house with new information that changes the child's behavior, as well as the family's. After speaking to a sixth-grade class about the importance of kindness to animals, a girl went home and after arming herself with information from "the Google," trashed every product in the house manufactured by companies that test on animals. I never discussed laboratory animals in my presentation, but the child made the connection herself. Her parents called and screamed at me, but then went out and bought cruelty-free products. This same scenario recurs constantly. Our children watch us carefully, and they love to point out how we screw up everything. Let's admit our mistakes, and help them fix things.

The Shelter Dog as the "It" Dog

The theory here is simple: raise awareness of existing dogs and pause the manufacture of new ones so that families and dogs can find each other and all breeds are able to replenish. If we love dogs enough to pay thousands of dollars for them and always keep them at our side, we should love them enough to not support irresponsible breeding. Whether the dog in the shelter is an "all-natural" mixed breed or a designer dog, he or she needs a home. With patience, not impulse, the dog for you is probably already here.

Efforts to raise this awareness are gaining momentum and popularity. In September 2015, the California State Legislature passed a concurrent resolution declaring the shelter pet as the official California State Pet. The resolution, ACR 56, was introduced by Assemblyman Eric Linder and read:

WHEREAS, There are currently around eight million abandoned pets living in animal shelters in the United States; and

WHEREAS, Three to four million of these dogs and cats are euthanized every year; and

WHEREAS, The Legislature seeks to raise public awareness of shelter animals; now, therefore, be it Resolved by the Assembly of the State of California, the Senate thereof concurring, that the Legislature hereby declares a shelter pet as the official State Pet.[59]

Flanked by those of us representing shelters, celebrities, and grateful consumers, Linder made a passionate plea for people to adopt from shelters and provide homes for existing dogs. The shelter pet joined California's thirty-seven other official symbols, ranging from the state flower (the golden poppy) to the state soil (San Joaquin soil). Georgia, Tennessee, and Colorado also recognize shelter pets this way. However, we were all bummed out when the California governor who signed this bill, Jerry Brown, purchased a designer dog from a breeder—a bordoodle (a poodle mixed with a border collie)—and not the state pet.

Another way to celebrate and raise awareness is National Mutt Day, also known National Mixed Breed Dog Day. This holiday was created in 2005 by "Celebrity Pet & Family Lifestyle Expert" and animal welfare advocate Colleen Paige and is celebrated on both July 31 and December 2. It is intended to promote awareness of mutts, many of whom live in shelters. The days are highly publicized on social media, with occasional sponsors directing funds to shelters through events organized for the same purpose. The purpose of National Mutt Day is described on the website:

---

59 California Legislature, "ACR-56 Official State Pet," Assembly Concurrent Resolution No. 56 (September 21, 2015), https://leginfo.legislature.ca.gov/faces/billTextClient.xhtml?bill_id=201520160ACR56.

National Mutt Day is all about embracing, saving, and celebrating mixed breed dogs. The biggest percentage of dogs abandoned and euthanized is due to the constant over-breeding and public desire of designer dogs and purebred puppies that are sold to pet stores supplied by puppy mills that often produce ill and horribly neglected animals.[60]

The "All-Natural" Designer Dog

The most interesting thing about mixed-breed dogs is that each is a one-of-a-kind dog. Part of the designer dog allure is to have something custom-made that nobody else has, something special. Yet, designer dogs are massed-produced to be identical to one another. Why are we forcing this? It's not like my yellow Lab/retriever/chow pines and mopes all day wishing he was a "real" dog, a pedigree. He does not beg to dress up as a French bulldog for Halloween. This makes no sense when there are truly singular and exceptional dogs looking for homes. Can we not embrace their differences? Unique, sometimes goofy-looking dogs need love too.

I would like to introduce a new breed: the all-natural designer dog. Often found in shelters, each of these dogs is a unique, natural, and organic designer dog. No pesticides, no manipulation, no forced errors, no miller, just a matchless product of nature. Full of hybrid vigor with all the sturdiness and purity of a first-generation breed, the shelter dog stands up to all and is bested by none. When interviewing a shelter dog, you see exactly what you get. The size, color, and coat are all present and accounted for.

At a shelter, the staff can tell you what they have seen of the dog's behavior and if the dog can do tricks, such as sitting, lying

---

60 National Mutt Day, http://www.nationalmuttday.com.

down, or giving you a high five. You can interact with the dog, bring your other dog to meet him or her, or just hang out and see if there's chemistry. Of course, there might be an illness or a behavioral issue, it's not a perfect system—everyone can have a problem—but for the most part, people love their all-natural designer shelter dogs.

It's important to bring folks to the realization that shelter dogs are great pets, so that they can make the informed and voluntary decision to adopt one. It's crazy to attempt to coerce people to adopt an all-natural dog by screaming at them that the dog might be destroyed within hours if it is not adopted. Doing this casts aspersions on shelters, is often defamatory since the statement is usually false, and in almost every case is the product of a self-serving agenda or fundraising scheme not disclosed. So, I ask you to ignore that noise and focus on the positive instead. Television stations around the country feature weekly adoption segments from local shelters. Watch them. Emulate the celebrities who adopt shelter pets. They include Betty White, Hilary Swank, Jane Lynch, Kaley Cuoco, George Clooney, Liam Hemsworth, Olivia Munn, Tom Brady, Amanda Seyfried, John Legend, Chelsea Handler, Andy Cohen, Bradley Cooper, Lena Dunham, Billy Joel, Zooey Deschanel, Miranda Lambert, Zac Efron, Miley Cyrus, Charlize Theron, Selena Gomez, Jake Gyllenhaal, and Jon Hamm, to name just a few. There are many more.

It's true that shelters don't always have puppies. But think about what a good deed it is to adopt a dog that's a bit older. They love you just as much, and puppies are a lot of work. Being patient and waiting for the shelter to get a puppy won't kill you either. If we at spcaLA know of another shelter that has puppies, we are happy to share that information. If you must have a puppy, ask your local shelter to do the same.

Goldendoodles, labradoodles, puggles, schnoodles (a cross between a schnauzer and a poodle), and the like are still mutts. Mutts, mongrels, mixed breeds, whatever you want to call them, they are mutts. While the designer mutts are doomed to suffer in puppy mills, pet shops, and backyard breeding basements, the all-natural dogs in the shelter are standing there asking someone to love them, someone who may be paying thousands of dollars for a mutt they think is a purebred. It is an absolute mockery of the shelter system for someone to say, "I have a purebred cockapoo."

If I sound like I am on the ledge over this, it is because I am. It is problematic enough that shelters already have to combat the myth that purebreds are more prestigious, healthy, and safe than shelter dogs (though 25 percent of shelter residents are in fact purebreds), but we now also have to compete with fake pedigrees who are real mutts and who still end up in shelters, where you can't tell them apart from the all-natural variety. You know by now that some of the shelter mutts are in fact "purebred" cockapoos that someone has already paid thousands of dollars for, prior to the dog arriving at the shelter. You can adopt that same dog for a fraction of the cost and use the savings to pamper him.

At one point, spcaLA introduced its own, exclusive designer dog breeds. At spcaLA shelters one can find the Chihuerrier (Chihuahua-terrier mix), the Yorkriever (Yorkshire terrier-golden retriever mix), the sheprador (German shepherd-Labrador mix), the houndranian (hound-Pomeranian mix), and the Chug (Chihuahua-pug mix). Exclusive new breeds emerge daily, and we can provide papers of "authenticity" with each adoption. The goal is to give the shelter dog a chance and showcase these all-natural, no miller required, pets.

It is worth mentioning that we are all paying the cost of the designer mutt compulsion and cruelty. Public shelters operate on tax dollars. You are paying to support all the excess, the discarded, ill, and behaviorally unsound dogs, animals who are products of impulse buying, trends, and cruelty. Would it not be better if the resources could be used to find a place to live for pets that never had a home, or use the resources to support pets already in families that may be experiencing hard times, or fund efforts to assist ethical breeders and researchers to find ways to undo the damage done to breeds by inbreeding, or other positive efforts that will offer health and a good quality of life for us all? Would you not prefer that? The cost of cute is doubled for every taxpayer who buys a designer dog, and those who don't buy them have to chip in as well.

The Painfully High Cost of Adorable

It goes without saying that there are hard costs in the designer dog industry. The dogs are priced anywhere from about $1,000 to $12,500, and that is just the sticker price. The dog could actually cost twice or three times as much if financed through a third party. In some cases, as in lease-to-own arrangements, the dog is not actually yours and you may find yourself in financial trouble that affects the rest of your lifestyle and possibly your credit score. As taxpayers, we all support the puppy mill industry, since tax funds support law enforcement, shelters, and government-funded laboratories. When you factor in the fact that dealers, bunchers, and street sellers deal in cash and that many sales transactions occur in cash, the cost is much higher. The sellers are not paying taxes on these sales, so the rest of us have less funding for law enforcement, infrastructure, and shelter resources.

Finally, we all pay another cost for cute. In the race to manufacture dogs as quickly and cheaply as possible for those who must have designer dogs, the puppy mill industry has actually destroyed those dogs and perhaps entire breeds for generations to come.

Even if the dog is healthy, other costs include routine medical necessities, food, treats, accessories, vacation boarding or travel, clothing if you are so inclined, grooming supplies, toys, and the like. Estimates average about $10,000 a year for a no-frills existence. Designer dogs who arrive with major medical problems requiring surgery or ongoing life-sustaining intervention due to congenital or hereditary defects can more than triple those annual estimates. One serious surgery can use up your entire annual budget for pet care. This presumes that the dog survives and lives a decade. If the dog does not survive, as is often the case, because of the horrendous start of his or her little life, then the cost includes not only the purchase price and medical costs, but also the grief, anger, and disappointment that is left behind for you to absorb. If the dog does survive but is not well, hard costs increase due to ongoing medical expenses. The quality of life enjoyed by you, your family, and your dog will be impacted.

It is not pleasant to watch an animal suffer daily. Watching a dog try to play and act like a normal dog but not be able to, with no idea why, is heart-wrenching. The dog loves you, you know that, but you can't fix the problem. So, what is the cost of the pain and suffering? What is the cost of your pain and suffering or the sacrifices you make to not leave the dog alone: to hire a pet sitter, to go home after work to tend to the dog rather than go out for drinks, to try to find a job where there's flexibility, so that you can give the dog injections or pills several times a day, to use vacation days from work to visit the veterinarian? What is that costing you? How about the ulcer you developed hating the person who sold you the dog who is now relaxing in the Cayman Islands where his bank is? Who pays for that? Do you have medical insurance? Perhaps you are suffering like this because you want your children to see that once you make a commitment, you honor it. That if the dog is sick, you take him to the doctor, just like you take them. But they don't understand why their dog does not act like their friend's dog, who catches a ball, goes on hikes, and horses around like a goof. So maybe now you need a second dog. Can you afford that?

Suppose you just get rid of the dog. First you might try to sell him. Do you disclose all the problems? If you do, there will be no sale. If you don't, you are no better than the person who duped you. What is the cost of your integrity, or the example you set for your children? Maybe you just let the dog out onto the street. He'll be okay, you might think. Dogs know how to cross streets, find food, and take care of themselves. No, you can't do that. The dog needs constant medical care and doesn't know the first thing about street life.

What is the cost of anguish? You talk to the veterinarian about euthanasia. The dog just doesn't have a good quality of life. You've never intentionally killed anyone before. Is the dog in agony or are you tired of the obligation? How would you explain that? Can you do it? There is a hard cost to the procedure and the veterinary visit. What is the cost of sorrow? So you give the dog to a shelter. You want assurance that the dog will be rehomed, but the shelter won't promise that, particularly given the condition of the dog. You are back to worrying about euthanasia. You berate yourself for being an idiot. You are a normal person; Madonna has a ton of help to care for her dog if needed. You are just you. What is the cost of feeling like a fool? You decide to leave the dog at the shelter. Grief-stricken and guilty, you say goodbye and leave. What is the cost of surrender?

Most of us would feel these emotions. People usually have a basically good ethical and moral core. If people know the implications of designer dog purchases *before* they make the purchase, they can avoid them.

It takes time to shift a way of thinking. Moving away from wearing fur as a fashion statement took more than a decade; the story of blood diamonds took time to circulate. But in the end, people heard the messages and sought alternatives. Most people would rather do no harm, and, given choices, will almost always pick the one that doesn't hurt. They will wear faux fur or buy a different kind of diamond. Given the choice to purchase a brand of aspirin, with a portion of proceeds donated to arthritis research, as opposed to one with no philanthropic gesture, they will pick the former. Look how we donate to fire and earthquake victims around the world. In general, I believe we really are a great group of humans.

As the word gets out about the pain and suffering that is caused by puppy mills, tormenting and exploiting animals for no good reason, people will stop wanting puppy mill dogs and the abuse will stop as the abusers search for a new get-rich-quick scheme. We can fix this.

A WINTER'S TALE

When someone tells you nothing is impossible, ask him to dribble a football.

—UNKNOWN

I have always been a large mutt person whose pets arrived serendipitously, unplanned, and unheralded. They've seldom been puppies, and they've always needed help, often the victims of animal cruelty. Several years ago, I became the mother of my first, and only, recovering puppy mill designer dog.

I was at one of spcaLA's pet adoption centers for a meeting when our shelter manager pulled me aside and said, "You have to see this." What I saw was the teeniest Pomeranian trembling in the corner of the kennel. She was filthy, matted, flea-infested, sick, wounded, and absolutely terrified. She was wandering the streets when she was fortunately found by animal control officers, who brought her in. There was no alternative but for me to take her home, foster her, and make her well so she could find a good home. She was simply too distraught to heal at the adoption center.

Once home, she would not let me touch her. She growled, snapped at my all-natural designer mutt, and was essentially Cujo disguised as a gross hairball. All I could do was sit, sob, and avoid being attacked. It finally dawned on me that between trying to survive on the street and not understanding the shelter environment, she probably had not slept peacefully in days, or longer. I put her in a covered bed next to mine, where she slept for forty-eight hours straight.

When she woke up, the change was miraculous. She was a new dog—calm, wagging her tail, and with no gnashing of teeth. We confirmed that she weighed one-and-a-half pounds (now, eight years later, she weighs four-and-a-half pounds), due to lack of nourishment, anemia, assorted worms, an intestinal infection, and lack of sleep. She also had open wounds, was too weak to sedate to deal with the matted hair, and was too ill to be spayed. It took about two months for Winter—it turned out that she was a white dog—to get well, become housebroken, and finally look like the Pomeranian she is.

To this day, she still demonstrates vestigial guarding and survival behaviors. For example, she fills her mouth with food and then runs to the other end of the house, spits the food out, and eats in secret. When finished with that mouthful, she repeats the process until she is finished. She does this even though we free-feed around the clock. She hides all her toys and treasures, including the things that she can snatch away from our other pets. Otherwise, she is friendly and has never snarled at anyone since the first night. I wish she could tell us what happened to her. She is too small to be a standard Pomeranian, too large to be considered a teacup, which makes her unsuitable to breed for show or anything else. She might have been the runt of a standard-sized litter or the giant of a teacup litter. We will never know.

Pomeranians are high-maintenance dogs who require a lot of hair-knot management and veterinary care. Some telltale classic inbred puppy mill surprises have already presented. She has a luxating patella (trick knee), collapsing trachea issues, occasional seizures, and trouble breathing—some of the most common problems puppy mill dogs have. She eats prescription food to help her joints. Caring for a dog like her is a significant expense. Often, the purchase price, exorbitant as it is, might be the cheapest part of this adventure. As the old saying goes, "The horse is cheaper than the oats." Still, I took her in, I made that commitment, she depends on me, and our bond is strong.

I must confess that after taking her in, I immediately went out and purchased beautiful dog purses for those occasions that I need to travel with her or keep her safely in the car. To be clear, the purses are specifically designed for carrying dogs. I do not toss her into my actual purse, though she would fit, to rattle around with all the items that I must have with me in case I am stranded on a desert island. Imagine a bad interaction with the dog, a fountain pen, and nail glue!

Ironically, now that she's all fixed up and spoiled to death, people criticize me for having what appears to be a purebred dog. They ask how in my line of work, I could do such a thing. I laugh and tell them how she came to be part of my family, and I remind them that about 25 percent of the animal shelter canine population are designer dogs at a shelter price tag.

So, after all these years, I have a designer dog. Instead of a superhero genius dog, I have a Nicole Richie purse dog. I should have named her "Lassie the Houndrainian" and called it a day.

ACKNOWLEDGMENTS

Without the help and encouragement of spcaLA staff, including but not limited to, Miriam Davenport, Ana Bustilloz, and Cesar Perea, there would be no files, photos, or research. In fact, there would be no book!

Additionally, as it became apparent that I could only think and write at my kitchen table, rather than in the many office spaces available to me, I must thank my family for simply eating over the sink during this process. I would also be remiss if I did not thank them for all the support they give me when I bring home dogs, cats, birds, reptiles, and tortoises who need a little extra care and a helping hand.

Finally, I would like to thank Dr. Phil for his brilliant foreword and those who without hesitation rallied to lend their names and support this effort.